I0123822

SOUNDS IN TRANSLATION:
intersections of music, technology and society

SOUNDS IN TRANSLATION:
intersections of music, technology and society

EDITED BY AMY CHAN & ALISTAIR NOBLE

ANU
THE AUSTRALIAN NATIONAL UNIVERSITY

E PRESS

E PRESS

Published by ANU E Press
The Australian National University
Canberra ACT 0200, Australia
Email: anuepress@anu.edu.au
This title is also available online at: http://epress.anu.edu.au/sounds_translation_citation.html

---

National Library of Australia
Cataloguing-in-Publication entry

Title:	Sounds in translation : intersections of music, technology and society / editor, Amy Chan, Alistair Noble.
ISBN:	9781921536540 (pbk.) 9781921536557 (pdf.)
Subjects:	Sound. Music--Acoustics and physics.
Other Authors/Contributors:	Chan, Amy. Noble, Alistair.
Dewey Number:	534

All rights reserved. No part of this publication may be reproduced, stored in a retrieval system or transmitted in any form or by any means, electronic, mechanical, photocopying or otherwise, without the prior permission of the publisher.

Cover design by ANU E Press

This edition © 2009 ANU E Press

Table of Contents

Acknowledgments

The editors would like to express their thanks and appreciation: firstly, to each of the contributors. We are grateful for their generosity in sharing their ideas, expertise and experience at the conference and later in this book, and for their patience during its production. We are particularly appreciative that they entrusted us with their works. We would also like to thank the ANU Faculty of Arts for funding the conference that this book is based on, and the ANU School of Music for providing resources for both the conference and the book. We are grateful to the referees for their invaluable comments and feedback on the manuscript. Lastly, we thank the ANU E Press, and especially Duncan Beard for the fantastic editorial work, advice and patience.

A. Chan and A. Noble
Canberra, 2009

Contributors

Freya Bailes

Freya Bailes is a postdoctoral fellow at MARCS Auditory Laboratories at the University of Western Sydney, where she combines her background in music and psychology using experimental techniques to investigate listeners' perceptions and emotional responses to music. Her principal research interests lie in the field of musical imagery, the imagined auditory experience of music. Her work has been presented in Europe, North America, Asia and Australia. She has published in book chapters and interdisciplinary journals including *Musicae Scientiae*, *Music Perception*, *Journal of New Music Research* and *Organised Sound*. Freya is an oboist who continues to perform whenever she can.

Amy Chan

Amy Chan received her PhD from The Australian National University, working on inter-cultural music and identity politics, with a focus on music from Malaysia and Singapore. Her research interests are in Western music in Asia, inter-culturalism and music, and world popular music. She lectures and teaches across a range of courses in performing arts and Asian studies. She currently holds a Visiting Fellowship at the Centre of Southeast Asian Studies at the The Australian National University. She co-convened the Sounds in Translation conference hosted by the Faculty of Arts at The Australian National University in 2005.

Jennifer Gall

Jennifer Gall holds a PhD from the School of Music, The Australian National University. The topic of her dissertation is 'Redefining the tradition: the role of women in the transmission and evolution of Australian folk music'. In 2006, she released the CD *Cantara*, which was based on her doctoral research. Gall has been involved with folk music as a performer, a collector, as a folk music consultant to the National Library of Australia and as guest lecturer for the ANU Folk and World Music courses. She currently holds the National Library of Australia and the National Folk Festival Fellowship for 2008–09. With Alistair Noble and Amy Chan, she was co-organiser of the 2005 Sounds in Translation conference.

Henry Johnson

Henry Johnson is Professor in the Department of Music, University of Otago. His teaching and research interests are in the field of ethnomusicology, particularly the music of Japan, Indonesia and India. His recent publications include *The Koto* (2004, Hotei), *Asia in the Making of New Zealand* (2006, Auckland University Press), which he co-edited with Brian Moloughney, and *Performing Japan* (2008, Global Oriental), which was co-edited with Jerry Jaffe.

Nicholas Ng

Nicholas Ng is a composer/performer/researcher based at the Queensland Conservatory at Griffith University. His doctoral dissertation, 'Celestial roots: the music of Sydney's Chinese, 1954–2004', integrated the disciplines of ethnomusicology and composition in the study of Australian Chinese sacred music in Sydney. Nicholas plays the *erhu* (Chinese 'violin') and has performed at venues including The Studio, Sydney Opera House, Yat-Sen Memorial Hall (Taipei) and Merkin Concert Hall (New York City), and is currently touring in William Yang's production *China*. As a composer, Nicholas has been commissioned by ensembles such as the Melbourne Symphony Orchestra, Foundation for Universal Sacred Music, The Song Company, The Australian Voices, United Nations Association of Australia, Australian Choreographic Centre, Tugpindulayaw Theatre, Sydney-Asia Pacific Film Festival and the Art Gallery of New South Wales. Published by Orpheus Music, his compositions have been broadcast on ABC Classic FM and awarded prizes such as the Orpheus Publications Composition Prize 2005.

Alistair Noble

Alistair Noble is a composer, pianist and musicologist based at The Australian National University. His music has been performed in Australia and America and notably recorded by Rotraud Schneider, Renata Turrini and Colin Noble. His current research includes study of Morton Feldman's compositional sketches. In 2005, he was co-convener of the conference Sounds in Translation, hosted by the Faculty of Arts, The Australian National University.

Alistair Riddell

Alistair Riddell studied music and computer science at La Trobe University in Australia and holds a PhD in music composition from Princeton University in the United States. He was a postdoctoral fellow at La Trobe University (1995–96) and president of the Australasian Computer Music Association (1994–96). His computer music works have been performed in New York, Brazil, London, Delphi, Berlin and Hong Kong. More recently, Alistair has been involved in interactive music performance and the development of kinetic installation art. He is currently a lecturer in sound art and physical computing in the School of Art at The Australian National University.

Phil Rose

Phil Rose is reader in phonetics and Chinese linguistics at The Australian National University and British Academy Visiting Professor at the Joseph Bell Centre for Forensic Statistics and Legal Reasoning at the University of Edinburgh. He conducts research into forensic speaker recognition and tones in Chinese dialects. He also undertakes FRS casework as a forensic phonetic consultant. His works have been published in journals such as *Journal of Chinese Linguistics* and

Australian Journal of Linguistics and in books, *The Technical Comparison of Forensic Voice Samples* and *Forensic Speaker Identification.*

Adam Shoemaker

Adam Shoemaker began at Monash University as Deputy Vice-Chancellor (Education) in September 2007. Before this appointment, he was Dean of the College of Arts and Social Sciences at The Australian National University and foundation director of the ANU Research School of Humanities. He holds a doctorate from The Australian National University. His sustained research interest is in Indigenous Australian history, literature, culture and politics, and he has been published extensively in these areas. He is currently co-writing a study of international Indigenous cultural flows called *Authenticity? Indigenous Culture and Globalisation.* Adam is active in community engagement and has held a variety of national and international appointments, most notably as chair of the Brisbane Writers Festival, president of the Association for Canadian Studies in Australia and New Zealand (ACSANZ) and president of the Australasian Council of Deans of Arts, Social Sciences and Humanities (DASSH).

Hazel Smith

Hazel Smith is a research professor in the Writing and Society Research Group at the University of Western Sydney. She is author of *The Writing Experiment: Strategies for innovative creative writing* (2005, Allen and Unwin) and *Hyperscapes in the Poetry of Frank O'Hara: Difference, homosexuality, topography* (2000, Liverpool University Press), and co-author of *Improvisation, Hypermedia and the Arts Since 1945* (1997, Harwood Academic). She is also a creative writer and has published three volumes of poetry, three CDs of performance work and numerous multimedia works. Her latest volume with accompanying CD-ROM is *The Erotics of Geography: Poetry, performance texts, new media works* (Tinfish Press, Hawai'i). Formerly a professional violinist, she is a member of austraLYSIS, the sound and intermedia arts group. Her web site is <www.australysis.com>

Introduction

Amy Chan

Sounds in Translation: Intersections of music, technology and society joins a growing number of publications taking up R. Murray Schafer's challenge to examine and to refocus attention on the sound dimensions of our human environment. His book *The Soundscape: Our sonic environment and the tuning of the world* (1977/1994) explores the idea of the 'sound-scape', the acoustic environment that we inhabit. Schafer invited researchers in the area of sound and music to investigate the origins, causes and impacts of new (and changing) sounds in rural and urban environments and to ask 'what is the relationship between man [sic] and the sounds of his environment and what happens when those sounds change?' (1977:3–4). He concluded that the 'home territory of soundscape studies' is at the crossroads of 'science, society and the arts' (Schafer 1977). Subsequent volumes have taken up this challenge: *Wireless Imagination: Sound, radio and the avant-garde* (Kahn and Whitehead 1992), *The Soundscape of Modernity: Architectural acoustics and the culture of listening in America, 1900–1933* (Thompson 2002), *The Audible Past: Cultural origins of sound reproduction* (Sterne 2003), *Sound Matters: Essays on the acoustics of modern German culture* (Alter and Koepnick 2004) and *Hearing Cultures: Essays on sound, listening and modernity* (Erlmann 2004). These books illuminate the sound aspects of a socio-historical period, be it in the Americas in the early twentieth century, in Germany in the past 200 years or in others. Their discussions also add to the current discourse on sound and the concepts of 'sound' and the sound-scape. As Alter and Koepnick have noted, there is still very little in our present critical vocabulary relating specifically to sound. The terminologies we presently use are terms borrowed from the vocabulary devised to describe the visual. *Sounds in Translation* seeks to add to this discourse and ultimately to expand this vocabulary to better interpret and explicate the actions and interactions of sound within domestic, public, urban, rural, private and performance environments.

The initial aim for this project was to open a dialogue between scholars and researchers in sound-related disciplines within the Asia-Pacific region. The conference that initiated the project drew researchers from various parts of Australia and New Zealand, discussing the multi-layered, multifaceted and multi-focal nature of sound. From the papers presented at the conference, we observed that there was certainly energetic research activity in sound studies from this region. Each of these articles responded to our call for papers on a topic that was inter/trans-disciplinary, crossing the boundaries of science, social science and the humanities. Discussion on sounds has been defined by the various disciplines: as a science subject, it has been the focus within areas of acoustics, sound production and technology (see works by Taylor 1965; Fletcher and Rossing 1998; Rossing and Fletcher 2004) and therapy (Unkefer 1990 and many more). In the field of social science, within the sub-disciplines of anthropology/ethnomusicology (for example, Feld 1982; Seeger 1987) and, most recently, sociology (as in the study of popular music) (Mitchell 1996; Hayward 1999 and many more), sound is studied as a manifestation of social meaning (as the expressed conduit of deeper social meaning), as traditionally adherent or otherwise. From a humanities perspective, sound is usually studied as 'music', and within Western culture and history, this has its own discourse and discursive practice. Needless to say, these studies do not easily share information across these boundaries and there is no established dialogic relationship between one study and another across disciplines. It was the objective of the conference and this book to provide the opportunity and space for such an interaction, particularly for those working within this geographical region of the Asia-Pacific.

In the process, we promote the interrogation of sound as a concept: how do its physical attributes (in terms of sound waves and its frequencies) translate into meaning? How have we imbued meaning onto sound objects, and what are their effects on human thought and activity? Can we understand sound outside of its music/social/cultural environment? When and how does noise cross over into music? When is sound 'art'? How is sound trans-cultural? How has the commodification (recording, packaging and selling/buying) of sound changed our perception of it? To answer these questions and more, we facilitated scholarly discussions on the theme 'sounds in translation', sharing ideas and practical experiments on the various facets of this concept. Our conference created a space in which to establish discourse and (hopefully) laid the foundations for further interdisciplinary exploration.

Sounds in Translation, the book, is our attempt to further consolidate what transpired during the conference and to allow amplified discussions and discourse on this topic. It is our hope that each strand of inquiry presented in this book will stimulate further investigations. The collection of essays reveals different facets of the concept/artefact of sound: as art (as in sound art) in the context of a staged performance, as a cultural practice bearing ritualistic significance, as enunciated in language and its day-to-day usage and as an identifiable object to be transferred from one person to another, one generation to another, one period to another. In each of these aspects, the concept 'sound' is conceived of in its full spectrum, from music to noise. What some might regard as noise (and therefore as 'waste matter' of human activity) is at the very least part of the sound-scape of this environment we live in. While some of our contributors discuss their manipulation of sound in various performances and contexts, others attempt to understand sound as it is.

The Sounds in Translation conference was conceived as a platform for discourse on sound (and the concept 'sound') outside the boundaries of musicology. Books such as Richard Leppert's *The Sight of Sound* (1993) and Andrew Ford's *Undue Noise* (2002) might appear to interrogate the idea of sound itself, but they are really appropriating the term 'sound' to discuss the cultural aspect of this sensory experience and, in most cases, the Western cultural aspect only. Music-centred analysis of sound—and here I'm referring to research and analysis undertaken within the disciplines of musicology, ethnomusicology and music theory—analyses and perceives sound through the prism of culture (and art and its practice as part of the larger cultural practice of any society). Academic 'discourse can become a site of struggle among the factions and interest groups that compete for the cultural authority to speak about music…[The] impulse to control and centralize scholarly production is forming discourse in the opposite direction…toward increasing uniformity' (Korsyn 2003:6). Korsyn goes on to discuss the identities scholars 'stake…on a particular mode of discourse'. This increasing uniformity has marginalised other works and research that are not immediately within its range and has discouraged the search for and recognition of other plausible prisms to perceive and conceive sound. It is our intention that this platform (in terms of the Sounds in Translation conference and book) is open to other prisms of analyses, perception and conception.

When we—the organisers and editors of the conference and this volume—looked beyond the familiar disciplinary area of music research, we found ourselves in the company of many unique researchers. Following Schafer's (1977/1994) notion of the 'interdiscipline' (as the ground or space where the three major branches of academia intersect), we chose our theme 'sound in translation' with the hope to further advance thinking on this interdisciplinary space. We use 'translation' for its connotation of *moving, crossing* and *process*. The word 'translation' can mean to *transfer*, to *transform* and as a *process* of expressing/expression (*The Oxford English Dictionary* 1978).

Using these various definitions, this book seeks to raise some salient questions regarding the translative process of sound: 1) what happens to sound during the process of transfer and transformation; and 2) what transpires in the process of sound production/expression/performance. Within the scope of performance (the expressed enunciated form of sound), we analyse the changes wrought—on either the sound component itself or its capacity for further signification. Critical appraisal of the process of translation, in particular its translation cross-culturally, is important to each of these essays. Hazel Smith's chapter ('"soundAFFECTs": translation, writing, new media, affect') discusses the combination of sound, image and text in a multimedia performance that is not music, literature, performance or technology, but all of the above. In the trans-media experiment of 'soundAFFECTs', Smith brings together multiple forms of media (sound, text and image) with the aid of technology provided by Roger Dean in the form of a real-time image-processing program. She then examines the process of translating the print-based text 'AFFECTions' to a multimedia performance of 'soundAFFECTs', and the changes in its affectability (on an audience) of each of these different performances. Alistair Riddell's *'Edible Audience*: what about this gastronomic performance translated as sound art?' is another combination of sound, image and technology, but with live performance that does not involve text. Riddell's chapter, like Smith's, examines the dual process of translation: it is, first, concerned with the translation of the act of dining into a sound-art performance, and second, its capacity to communicate to a general audience, in particular how an audience will be able to construct meaning from a non-verbal technology-based performance.

Freya Bailes' chapter discusses the losses and gains that take place in the process of transfer between the point of input, when a musician reads a musical score, and the point of output, when the score is realised in music during performance. With her case studies, she demonstrates that in the process of translating sound to mental imagery the 'veridical nature of sound colour' is lost, but the re-imaging through active imagination augments its expressivity and meaning. Nicholas Ng's chapter, 'Domesticating the foreign', features a similar process of loss and gain, highlighting also a dual-translation process: one of language and linguistic differences between Mandarin, Cantonese and English, and another of the transference of Catholic practice from Rome (originally), via Hong Kong/Taiwan and finally to Sydney. Both processes led to a negotiated compromise: a hybrid of language and practices as accumulated through the migratory and assimilative processes in which the 'purity' of a single language or practice was rendered unimportant.

Jennifer Gall's chapter plays with ideas of impurity, or inauthenticity, by examining the incorporation of the sound-scape of a recording of a song made in the 1950s into her reproduction of the same song in more recent times. Her discussion (and performance) of *Green Bushes* recognises the 'inauthenticity' of her version of the folk song and, in fact, plays it up. While on one hand, Gall's performance attempts to retain the identity of the folk song along with its recorded context (that is, the kitchen), her 'play' on its authenticity, as reflected in the digital manipulation provided by Ian Blake, demonstrates its impossibility. Her performance (as argued in her chapter) reflects the interaction (the to and fro) between the past and the present and the impossibility of recreating a replica in a present-day performance of the song. Adam Shoemaker's chapter on the sonic reverberations of the archaeological discovery of the 'footprint site' at Lake Mungo in New South Wales similarly traces a linearity between the past and the present, questioning the sonic busyness of the place/space on the basis of a personal, experiential understanding of the site. He mixes the present-day sound of digging, camera shots and footsteps (of the many researchers and journalists) with the long-ago (imagined) sound of the dead as conjured by this discovery.

My own chapter aims to question sound's meaning and capacity to represent in a changing environment, using as a case study a traditional drum performance as restaged on the concert platform. I argue that the translation of the *shigu* (Chinese drums) from one performance context to another (from a traditional environment to a modern/artistic one, each with its own values and significance) has, first, de-territorialised, and second, re-territorialised it, altering its nature, identity, value and representation in the process.

Phil Rose's 'Singing the syllables' and Henry Johnson's 'Voice-scapes: transl(oc)ating the performed voice in ethnomusicology' find their speaking positions between language and music. Rose and Johnson demonstrate the proximity of sound and speech in their respective chapters. While Rose raises the plausibility of the use of musical techniques and analysis to shed further light on Tibetan spelling chant, Johnson argues that speech and song are not separate entities but are better conceived as two poles of a continuum. The cross-disciplinary approach of linguistics and music allows for a more nuanced understanding of sound and sound products as they appear within the sound-scape.

Sound is ephemeral and ubiquitous or, as Alter and Koepnick (2004) put it, 'pervasive, invasive and evasive'. As a subject, it is difficult and problematic to represent and has not been, until recently with the aid of technology, able to 'speak for itself'. The audible sound often required the visual/text (be it in notation and graphs or in language) for its representation, conservation and transmission. It is only in recent times, with the aid of technology, that we have been able to 'capture' sound and transfer it. More recent technology, however, has allowed for almost instantaneous transmission across spatial distances via the Internet. It has enabled sound to represent itself, albeit mediated by technology. *Sounds in Translation*, the book, was conceived to take advantage of this new technology and the development in publishing of the electronic book. Much of what is written in the book is best illustrated by the sound itself and, in that sense, permitting sound to 'speak for itself', as none of the volumes mentioned earlier has that capability. We hope that the simultaneous accessibility of the book in print and e-format will provide better and easier access for our readers, and also engender a more instantaneous discussion and feedback.

Bibliography

Alter, Nora and Koepnick, Lutz (eds) 2004, *Sound Matters: Essays on the acoustics of modern German culture*, Berghahn Books, New York.

Altman, Rick (ed.) 1992, *Sound Theory: Sound practice*, Routledge, New York.

Erlmann, Veit (ed.) 2004, *Hearing Cultures: Essays on sound, listening and modernity*, Berg, Oxford.

Feld, Steven 1982, *Sound and Sentiment: Birds, weeping, poetics, and song in Kaluli expression*, University of Pennsylvania Press, Philadelphia.

Fletcher, Neville H. and Rossing, Thomas D. 1998, *The Physics of Musical Instruments*, Second edition, Springer Verlag, New York.

Ford, Andrew 2002, *Undue Noise: Words about music*, ABC Books, Sydney.

Hayward, Philip 1999, *Widening the Horizon: Exoticism in post-war popular music*, John Libbey, Sydney.

Kahn, Douglas and Whitehead, Gregory (eds) 1992, *Wireless Imagination: Sound, radio and the avant-garde*, MIT Press, Cambridge, Massachusetts.

Korsyn, Kevin 2003, *Decentering Music: A critique of contemporary musical research*, Oxford University Press, Oxford.

Leppert, Richard 1993, *The Sight of Sound: Music, representation, and the history of the body*, University of California Press, Berkeley.

Mitchell, Tony 1996, *Popular Music and Local Identity: Rock, pop, and rap in Europe and Oceania*, Leicester University Press, London.

Rossing, Thomas D. and Fletcher, Neville H. 2004, *Principles of Vibration and Sound*, Springer-Verlag, New York.

Schafer, R. Murray (ed.) 1977, *Five Village Soundscapes*, The Music of the Environment Series, no. 4, A. R. C. Publications, Vancouver.

Schafer, R. Murray (ed.) 1977/1994, *The Soundscape: Our sonic environment and the tuning of the world*, Destiny Books, Rochester, Vt.

Seeger, Anthony 1987, *Why Suyá Sing: A musical anthropology of an Amazonian people*, Cambridge University Press, Cambridge.

Sterne, Jonathan 2003, *The Audible Past: Cultural origins of sound reproduction*, Duke University Press, Durham.

Taylor, Charles 1965, *The Physics of Musical sounds*, E. U. P., London.

The Oxford English Dictionary 1978, Volume XI, Clarendon Press, Oxford.

Thompson, Emily 2002, *The Soundscape of Modernity: Architectural acoustics and the culture of listening in America, 1900–1933*, MIT Press, Cambridge, Massachusetts.

Unkefer, Robert F. (ed.) 1990, *Music Therapy in the Treatment of Adults with Mental Disorders: Theoretical bases and clinical interventions*, Schirmer Books, New York.

Wallin, Nils L., Merker, Bjorn and Brown, Steven (eds) 2000, *The Origins of Music*, MIT Press, Cambridge, Massachusetts.

1. 'soundAFFECTs': translation, writing, new media, affect[1]

Hazel Smith

This chapter will focus on the translation or trans-coding of a work from its print form as 'AFFECTions: friendship, community, bodies' (Brewster and Smith 2003) into its multimedia form as 'soundAFFECTs' (Dean et al. 2004).[2] In particular, it will explore how this transition changes the affective experience the piece transmits. The print (words only) work 'AFFECTions' is an experimental and multi-genre collaboration by Anne Brewster and myself, which engages with the subject of affect, feeling and emotion (Brewster and Smith 2003). The multimedia work 'soundAFFECTs' employs the text of 'AFFECTions' as its base, but converts it into a piece that combines text as moving image and transforming sound. For the multimedia work, Roger Dean programmed a performing interface using the real-time image-processing program Jitter; he also programmed a performing interface in MAX/MSP to enable algorithmic generation of the sound. This multimedia work has been shown in performance on many occasions, projected on a large screen, with live music; the text and sound are processed in real time and each performance is different.[3]

These translations/trans-codings[4] of the piece are all radically different creative enactments of it and the multimedia work—because it is processed in real time—is itself variable each time the work is encountered. The different versions are therefore not translations in the sense of original and copy, even though the print version was written first. In his essay 'The task of the translator', Walter Benjamin (1999) suggests that a translation is much more than the reproduction of meaning; rather, it is a creative reworking of it. For Benjamin (1999:81), the translator is in error 'if he preserves the state in which his own language happens to be instead of allowing his language to be powerfully affected by the foreign tongue'. Tim Woods (2002:200) argues, very suggestively, that in Benjamin's essay translation is seen as a kind of de-familiarisation, an 'eruption of the foreign' that foregrounds the idea that

> a text is never an organic, unified whole…A translation is therefore not simply a departure from the original that is either violent or faithful, since the original is already divided, exiled from itself. Not only is no text ever written in a single language, but each language is itself fractured. (Woods 2002:201)

The idea of translation as transformative production can also be applied to the trans-coding of works into different media. In such trans-codings, mutation is more important than fidelity to an original, and one version is not better or truer than another. Comparison of the different versions produces difference as much as similarity, and such differences are likely to already be potential in the prior text.

I want to explore what happens when the verbal text 'AFFECTions' transmutes into the multimedia work 'soundAFFECTs'. In particular, I want to focus on the way in which the emotional/affective aspect of the print version changes in the multimedia translation as a result of technological intervention and the fusion of text, image and sound. In order to do this, I will need to distinguish my use of the terms 'emotion' and 'affect'. I will define emotion as subjectively based but culturally coded categories of feeling such as happiness and anger, and affect or affective intensities as sensations, or a flux of sensations, less tied to a particular point of view or subjectivity.[5] My argument will be that there is a shift in the multimedia version so that it communicates much more strongly through affective intensities than the print version, and much less through the depiction and communication of subjectively based emotional states. The potential for this shift is, however, inherent in the print version, which oscillates between emotion and affect as I define it here and is therefore already marked by difference—'divided, exiled from itself' (Woods 2002:201). An important part of my discussion will be exploring theoretical frameworks from cultural theory and cognitive psychology—discourses that were themselves influential in the writing of the piece—as a means to consider the change in emotional/affective experience that takes place in the trans-coding from print to multimedia.

The print version: emotion and affect

In contemporary literary studies there has been a problem with regard to the discourse about emotion of which critics are becoming increasingly aware.[6] In the 1970s or 1980s, literary studies—appropriately and very importantly—problematised the humanist assumptions underlying notions of writing as self-expression and foregrounded the mediating activity of language. As a consequence, it somewhat bracketed out emotion/affect. For literary studies to reincorporate emotion it needs to distinguish, as Rei Terada does, between emotion and self-expression. In order to do this, Terada (2001) argues, somewhat confusingly, that emotion is non-subjective. It is also necessary to discuss emotion in literature beyond narrowly equating it with certain kinds of realist or lyric writing in which codifiable emotions are readily recognisable, and which communicate through realist strategies of emotional involvement and identification. In some forms of experimental poetry, for example, in which realist representation and a cohesive subjectivity are problematised, emotion is less subjectively based and is at least partly broken down into affective intensities. I want to show how 'AFFECTions', the print-based collaboration from which the multimedia work was made, oscillates between emotion and affect, so that the affective intensities accentuated in the trans-coded/translated multimedia version are already partly present within it. In order to do this, I need to retrospectively talk about the writing of 'AFFECTions' and its theoretical bases.

'AFFECTions' is a ficto-critical, mixed-genre work. Ficto-criticism is a form of writing that brings creative and theoretical writing into a resonant relationship with each other and creates symbiosis and friction between the two (Brewster 1996; Kerr and Nettelbeck 1998). The piece is therefore a mixture of fiction, poetry, theoretical exposition and quotation. It consists of about 30 short sections, or modules, with varying degrees of disjuncture and conjuncture between sections. 'AFFECTions' is a constellation of short narratives, poems and sections of theory, but there are no through narratives, consistent metaphors or overriding arguments. Rather the piece explores affect from a number of different perspectives without trying to resolve them. The critical and creative works are juxtaposed, but one does not explain or illustrate the other; rather they exist in a reflexive, porous and open relationship. The piece does not take a position on affect, but circles round the topic in ways that bring together diverse, even conflicting, perspectives. It includes interjections about affect in relation to the process of writing, performance, the media, the war in Iraq, cyberspace, music, dance, ethics, representation and so on.

In order to create the piece, we read widely within the literature about affect and emotion—though not systematically or comprehensively. In other words, our reading was exploratory, and we read to trigger creative responses, rather than to produce an overarching intellectual argument. Our reading was drawn from a number of different fields including literary and cultural theory, philosophy and cognitive psychology, but in retrospect two theoretical perspectives seem to hover over the piece. These two perspectives are, broadly speaking, from cognitive psychology and cultural studies. They also form the basis of my analysis, though this includes material I have read since we wrote the piece as well as before, and I will not attempt to distinguish between the various stages of that reading. Whereas the initial reading by both of us provided a trigger for the creation of the piece and the ideas in it, further perusal of the relevant theory has helped me to ponder retrospectively how the piece talks about affect and encodes its own affective experiences. It has enabled me to conceptualise how this emotional/affective experience is transmuted when the piece is trans-coded or 'translated' from print to multimedia.

The first of these perspectives stems from the work of cognitive psychologist Keith Oatley and is geared largely to emotion as I have defined it above. Oatley argues that emotions are cognitive responses, accompanied by bodily sensations, to our tendency to make plans and goals. Drawing on his work with Johnson-Laird, Oatley (1992:46) proposes that 'an emotion occurs in relation to a person's several plans and goals when there is a significant change in assessment of the outcome of a goal or plan'. The matter is, however, complex because we have multiple goals that are in conflict with each other and emotions enable us to coordinate these goals. We experience positive emotions when this coordination is successful, negative emotions when it fails (Oatley 1992). For Oatley:

> [E]motions derive from the cognitive processes for integrating multiple and sometimes vague goals and for managing plans that are enacted with limited resources in an uncertain environment, often in conjunction with other people. Happy emotions occur when coordination between plans is being achieved and unanticipated events are assimilated. Distressing emotions occur when coordination fails, or when some plan goes badly, when a problem emerges that cannot be solved from current resources or when an important background goal is violated. Emotions function to allow otherwise disparate aspects of a complex system to be co-ordinated. (Oatley 1992:43–4)

According to Oatley, therefore, emotions serve useful functions in helping us integrate our goals and plans, and can be important for quick decision making because we do not have to sift through all the arguments or possibilities in the way that might be necessary to make that decision by purely logical means. Oatley questions the idea that emotions are necessarily irrational while thought is rational—even if that is true in some cases—and views emotions as aids to thinking and behaving. Emotions can be aids to making decisions rapidly in cases where there is incomplete information: 'What they do is prompt us in a way that on average, during the course of evolution and assisted by our own development, has been better either than simply acting randomly or than becoming lost in thought trying to calculate the best possible action' (Oatley and Jenkins 1996:258). This, according to Oatley and Jenkins (1996), is an example of heuristics; a heuristic is 'a method of doing something that is usually useful when there is no guaranteed solution'.

Oatley is unusual in the way he draws many of his examples from literary texts, blurring the distinction between real life and fictional cases. These are, however, usually realist nineteenth-century texts such as *Anna Karenina* or *Middlemarch*, in which the emphasis is on character and situation. Oatley tends to analyse situations within the novels in terms of the emotions the characters experience and how these relate to the frustration or fulfilment of their plans. He does not attempt to look at affect in less realist fiction or poetry and he also does not consider the psychology of the reader and how this might be involved in the means by which the work communicates. In particular, he does not attempt to investigate how the reading process itself might be characterised by emotional interruption or fulfilment.

On the other hand, our reading also encompassed cultural studies material that was influenced by the work of Deleuze and Guattari (1994). This material tends to interrogate the assumptions behind more humanist perspectives and consequently is less cognitively and subject/person based. Rather it stresses pre-personal and non-subjective 'intensities'. For Deleuze and Guattari, everything belongs to a flow of 'becoming', which constitutes one immanent plane of being. According to them, we have to perceptually and intellectually carve up the world to understand it, but fundamentally everything is interconnected. Deleuze and Guattari do not deny the presence of the subject, but see the subject as an artificial construction. For Deleuze and Guattari, emotions and perceptions become affects and percepts at least partially detached from a point of view centred in a specific subject. Here the distinction between subject and object central to cognitive theory collapses, as affects cross over, engage with and move between human and non-human bodies producing 'affective intensities'. Deleuze and Guattari also draw attention to different formations of sensations such as 'the embrace' or 'the clinch' (the coupling of sensations) and 'withdrawal, division and distension' (the uncoupling of sensations). Affect is closely linked in Deleuze and Guattari's work with transformation and even with creativity itself. They suggest that artists are 'presenters of affects, the inventors and creators of affect' and that 'a great novelist is above all an artist who invents unknown or unrecognized affects and brings them to light as the becoming of his characters' (Deleuze and Guattari 1994:174).

We also explored cultural studies material, particularly the work of Brian Massumi (2002), which focused on the relationship between affect and politics: the way in which responses to political events could be primarily affective and the means by which affect was manipulated through the media by politicians to draw the populace into line with a conservative view of current events.

Both these perspectives (the cognitive and the cultural) are incorporated into the collaboration in direct and subliminal ways. In some sections, the theoretical input is quite direct, such as in the following passage, which outlines Oatley's ideas and then moves outward with the idea of the plot—the fictional equivalent of Oatley's 'real-life' plan:

> Keith Oatley argues that emotions are cognitive processes that arise from our tendency as human agents to make plans, rather like plots in a narrative. Positive emotions occur when goals are fulfilled, negative ones when they are thwarted. The situation is complex, because we usually

have multiple conflicting goals; our plans involve other people; and we often have to make decisions in an unpredictable environment.

Not just one plot, then, but plots within plots, competing plots, and plots without beginning or end. (Brewster and Smith 2003)

Another passage also draws on theoretical material from Derrida, and implicitly Massumi. It introduces the notion of the manipulation of feeling by politicians at the start of the Iraq war, partly through the promotion of stereotypical and negatively geared emotions towards the 'enemy'. The affective reaction of the speaker is one of bodily dysfunction:

It's 11.30 p.m., just before Bush is to address the US. I am almost incapacitated by a tired sick feeling. I guess we have all become drugged by the American imperative to feel—outrage, fear, pride (and our intense counter-feeling which John Howard named this morning in his address to the nation as 'rancour') which started with the attacks of September 11th. We watched with a growing frustration the incitement in the US of a powerful discourse of feelings, an instrumentalist military sublime. The American president drew a justification for war-mongering on the basis of his feelings of outrage. He once again arrogates to the American people the right to be human, to suffer; correspondingly the inhuman is returned to the third world, which has no claim to a collective subjectivity.

A journalist asked John Howard this morning whether he saw historical precedents (such as Viet Nam) in the current situation and he blithely dismissed 'history'. At times like this, he said, we can only think about the present. This is precisely where a discourse of feelings is so politically expeditious; it erases the history of antagonisms and an analysis of causes (such as a rapacious US foreign policy). And so we see the insidious effects of instrumentalist feeling in what Derrida calls the grotesque 'onto-theology of national humanism'.[7] (Brewster and Smith 2003)

Different approaches to emotion and affect hover over the piece thematically, but formal aspects of the piece can also be construed in a similar light. On the one hand, there is more subject-based writing—that is, writing where we are aware of authors, characters, narrators or voices as focal points, even though they are highly constructed and tend to convey emotions that are conflicting, complex and destabilising. These passages tend more towards the depiction of emotional situations and the pressure towards emotional identification by the reader. On the other hand, the piece also includes types of writing that are less subject based and that convey affect in a way less tied to a particular point of view or focalisation. These passages move closer to the notion of affective intensities. Obviously, these two extremes (the more subject based and the less subject based) form the end points of a continuum and most of the writing is along that continuum rather than at its extremes.

Similarly, the collaboration involves different ways of engaging with affect and emotion through linguistic and generic strategies aimed at representation or the breakdown of representation. On the one hand, it consists of more (though by no means entirely) narrative, realist and expositional types of writing, which at least partially encourage the illusion of emotional identification by the reader with situations and characters within the text—as well as some passages that directly transmit theoretical ideas. At the same time, it also includes types of writing that break up semantic, narrative or descriptive continuities. These types of writing tend to disrupt emotional categories and dissolve them into a flow, or assemblage, of sensations/impressions or 'affective intensities'. Such writing has a strong connection with the long tradition of twentieth and twenty-first century experimental writing and, most recently, with American language writing and its various successors.

Two examples will serve to illustrate the way the different modes of writing are inscribed in the text:[8]

I wonder if I am still alive at UNSW
the dead are remembered
but those who remember them
die also

The light is on in my room and
someone else is opening
my filing cabinet
which probably has finally collapsed

That last day we took photos hoping to hold on to
that which must inevitably slip away
we grin into the camera
stiffening to a trace

Meanwhile my email address passes over into redirection

You're glad to have escaped
but you'd also still like to be imprisoned
because it's safe

Sometimes I'm sad
sometimes I claim to be sad
sometimes I say I'm not sad but I am
I don't recognise a feeling until I'm falling out of it

Everyone thinks that their experience is singular
but everything is at least double
and we all know
that memories have limited respect
for identity

The best workplace is a mobile home
and moving through means passing on
but every day your ghost kicks into me
UNSW
(Brewster and Smith 2003)

had until today
whether or
not as revealed last week a
grave decision until yesterday
declared the skies surround
his strongholds
dailytelegraph.com immediately
defiant talk coaxed
shock and awe
in an hour appointed
brandishing a
tens of thousands a
political pause flammable mist
stunned
into under cover
whether or not
or not or not or
(Brewster and Smith 2003)

In the first poem, the sense of subjectivity is fragile. The speaker wonders if she can still be alive if she is not remembered and expresses considerable doubts about what her feelings really are: whether they are joy at 'escaping' or regret at no longer being in the former workplace. Feelings are not characterised as being easily coordinated cognitively: 'I don't recognise a feeling until I'm falling out of it'. Despite the fragility of the subject, however, and her lack of emotional and cognitive coordination, there is a sense of a focal subjectivity and an identifiable situation: the removal from one work environment to another. In Oatley's terms, the emotional ambivalence is caused by the conflict between two goals: the desire for change and a new workplace, and nostalgia for the previous one.

In the second piece, however, an overall point of view, or single identifiable situation, is less present. This is hardly surprising because the piece is a collage composed of fragments taken from newspaper headlines during the (second) 2003 Iraq invasion by the US-led coalition. In this piece, there is less sense that the fragments project a unified point of view. The indecisiveness or hesitancy at the end is not that of a particular person and, while the fragments suggest a situation (for example, 'shock and awe' was the name given to the attacks in the Iraq war), that situation is conveyed in a way that is indeterminate and fractured. It does not relate the sensations it transmits—such as stunned—to a particular consciousness, or locate them in a personalised emotional state. Furthermore, different viewpoints—such as that of the propagandist government and of the reader of the newspaper—are collapsed into each other, so that an overriding viewpoint is splintered and multiple perspectives unfold, which include feelings of disarray and impotence. These are not firmly rooted in a consistent subjectivity and as a result seem close to the concept of affective intensities.[9]

Overall, then, the print version oscillates between emotional identification and affective intensities, taking up also various positions in between and transmitting ideas about emotion and affect through its content and its form.

From print to multimedia and from emotion to affect

I now want to explore how in the translation/trans-coding from print to multimedia the piece shifts more strongly from emotional identification to affective intensities. The multimedia form of the piece is called 'soundAFFECTs' (Dean et al. 2004) and it brings together text, image and sound. 'soundAFFECTs' uses 'AFFECTions' as its base though it does not employ the whole text. Sections of the text now take the form of modules, each created as still or moving frames of a digital video, which are rearranged in each digital version. Roger Dean processed the modules, with programs written by him within the real-time image-processing platform Jitter, treating the text as a series of visual objects. The most important difference between the multimedia version and the page-based version is its variability. In the multimedia version, there is no fixed text; the order of texts and the way any particular one is processed will be reconstituted each time. The soundtrack, or at least a significant portion of it, is generated by Dean from the same algorithms and is also variable. This means there will always be some shared features between the versions but also considerable differences. So this is a form of creative production that is dynamic, productive and generative, unlike the print text, which remains identical with itself on a material level even if it is composed of differences at the level of content and style.

A number of different processes occur in this multimedia version. The texts are treated as visual objects in blocks. They are superimposed on each other; they are also stretched or compressed. Texts disintegrate into, and overwrite, other texts (though they do not necessarily overwrite them in the sense of replacing them). Texts are repeated; the screen divides into several sections, sometimes multiplying the same text, sometimes combining different ones, and so on. At times—particularly towards the end of the processing—the text becomes an intensively visual, dynamic and kinetic spectacle with the screen divided into several segments and a number of different processes operating at once with considerable rapidity. The multimedia version of the piece greatly accentuates certain characteristics already inherent in the print text towards interruption, fragmentation, circularity and non-linearity. It also speeds everything up enormously, creating a sense of extreme flux. The processing and the speed problematise the reading process: they are 'flickering signifiers' (Hayles 1999:47–8). Sometimes the words disappear before they can be fully digested; sometimes they appear in only partly readable forms or even in forms that can hardly be read at all. The multimedia version speeds reading up (we cannot read at our own pace and must scan the text much of the time rather than reading it); it also promotes movement and transformation of the text.

Most relevantly to my argument, the multimedia version breaks down the semantic, narrative, expositional aspects of the print version. In so doing it largely erodes a sense of the subject (distinct authors, narrators, characters or voices). The words point momentarily to authors or narrators who themselves flicker, transform and dissolve. Relevant here is Katherine Hayles' idea that the binary opposition between presence and absence (so predominant in some post-structuralist thinking) has been replaced in the discourses of informational systems by the binary of pattern and randomness. Hayles (1999) also argues, however, that the need to theorise embodiment as part of the cyber experience means that pattern must go hand in hand with presence and randomness with absence. In 'soundAFFECTs', words become patterns not only because they transform into visual designs, but because they convey shifting patterns of meaning that are not continuous or sustained. At the same time, the ghostly sense of voices/bodies that put these patterns into motion, or arise out of them, continuously haunts the text. These voices/bodies are those of the authors who are writing the text, the narrators who transmit it and the characters who inhabit it.

This leads us to the multi-sensory aspect of the piece: text, image and word combine and this greatly increases the range and interaction of sensations. An important aspect of this is what I have elsewhere called semiotic exchange—that is, the way different media, when brought together, can modify, even take on, each other's characteristics (Smith and Dean 1997). In 'soundAFFECTs', the text becomes image, while the sound intensifies the content of the words. This is a type of synaesthesia—that is, one sensory modality is 'translated' into another so that the meaning of the words is experienced in terms of sonic and visual stimuli (in poetry, synaesthesia usually alludes to the transference that occurs in metaphor from one sensory realm to another). Synaesthesia is central to multimedia practice, but in 'soundAFFECTs' it also takes the form of what we have conceptualised as 'algorithmic synaesthesia' (Dean et al. 2006). In algorithmic synaesthesia, image and sound share, at least partly, the same data and algorithmic processes. In 'soundAFFECTs', for example, many of the sonic effects are 'translations' of the movements of the text as image, and in the web version of the piece it is noticeable that the greatest density of sound is where there is most movement in the image. In addition, some of the source sounds that can be implemented in renderings of the piece have been made using programs such as Metasynth, in which a static image is read kinetically by an algorithm that generates sound.[10]

The sound, in particular, is extremely important in the shift from emotion to affect because sound (in general) *is* a flux of sensations, even more than words. Words, in contrast, always bear the burden of a referentiality that partly interrupts and fixes the flux and point to concrete situations in which pure sensation is dampened, objectified and solidified. Discussion of emotion in music—for example, in the work of Meyer—has often centred on the way it evokes emotional states deriving from the fulfilment or frustration of expectations (see Meyer 1956). This is somewhat akin to Oatley's concept that emotions arise in response to the fulfilment or interruption of goals. The concept, however, of a flux of sensations—of 'affective intensities'—seems to address more directly music's abstract, less cognitive and less referential aspects. This is particularly significant with regard to computer music, in which audience expectations are likely to be less pronounced than in more traditional musical forms.

In '*soundAFFECTs*', the sound continually moves the words beyond the domain of the referential while also interacting with it. Lawrence Kramer's (2002) work on mixed media and musical meaning is relevant here. He argues that music is more 'semantically absorptive' than other media. When juxtaposed with the 'imagetext'—a term he has adopted from W. J. T. Mitchell to signify the fusion of image and text—music takes on the semantic meanings generated by that imagetext. At the same time, musical meaning exceeds the referentiality of the imagetext. In mixed media, Kramer (2002:153) suggests, meaning 'runs on a loop'. The music seems to emit a meaning that 'it actually returns, and what it returns, it enriches and transforms' (Kramer 2002). For Kramer:

> From the standpoint of the imagetext, music has greater communicative immediacy, though less communicative power. Music, indeed, is one of the defining modes of an immediacy that the imagetext has to exclude in order to stabilise itself, to enable its generalising, abstracting, and speculative capacities, even at the cost of an ambivalent fascination with the excluded and excluding other. But as soon as meaning effectively runs from the imagetext to music along the semantic loop, the music

seems to convey that meaning to and through the imagetext in preconceptual, prerepresentational form. (Kramer 2002:153)

While Kramer does not refer specifically to affect here, the affective properties of music—and I mean affect here rather than emotion—seem to be implied in this idea of the capacity of music to return meaning to the imagetext in 'pre-conceptual, pre-representational form'.

The sound in 'soundAFFECTs' is generated algorithmically and is different in each performance. Nevertheless, all renderings of it will involve the genre known as noise: this means that the sound changes, but the changes come from relative distribution of energy in the frequencies, rather than abrupt transitions of pitches and 'notes'. There is, therefore, a continuous flow of sound with some variation, but it is not structured in a way that segments it or emphasises beginning, middle and end. It is rather like Deleuze and Guattari's (1987:22) description of a plateau as 'a continuous, self-vibrating region of intensities whose development avoids any orientation toward a culmination point or external end'. In several sequences in the web version, for example, the sound consists of a band of frequencies that moves from low to high in a repeating cycle. The impression when listening to the opening of this version of the piece is of protracted ascents of rising pitches. These slow ascents then develop into simultaneous ascending cycles at different speeds but with increasing density and levelling out of the pitch as the piece progresses. In addition, the use of multi-channel spatialised sound when the work is performed live produces a high degree of immersion for the audience, which helps them to receive the work with immediacy and as a flux of sensations. This immersion is greatly accentuated by the darkened room, the large screen and the sharing of the experience with others. In this respect, the experience of the web version is different from that of the performance version(s).

All these factors result in a much stronger push and pull between continuity and discontinuity in the multimedia version than in the print version. On the one hand, the texts are more interruptive in the multimedia than the print version and rapidly displace each other; there are also interruptions when from time to time a black screen lingers between texts. In other respects, however, the multimedia version is much more continuous than the print version: there are no gaps between the texts, except for the occasional black spaces, there is considerable repetition of texts and the disparate texts are welded together with image and sound. The overall effect is that the multimedia version breaks up the emotional ups and downs and emotional identifications that characterise some sections of the print version into a flux of sensations. As a result, it produces an affective environment, which fluctuates more continuously than the print version.

The question remains, then, how we can theorise this change in affective experience in the trans-coding from print to multimedia. I suggest that in order to do this we look at it through the lens of the cultural and cognitive theory I mentioned earlier. In terms of Deleuzian theory, the multimedia version is stronger, as I have already implied, at creating 'affective intensities'. Sensations couple and uncouple in the way Deleuze and Guattari describe, as the texts are superimposed on each other or disintegrate, and the words couple and uncouple with image and sound. The multimedia version increases, much more than the print version, the flow of becoming (that is, it speeds everything up). Deleuze talks about how we slow the flow of becoming, the vast and chaotic data we receive, in order to perceive and comprehend the world. Speeding it up again is perhaps to return, even if momentarily, to the state of flux and becoming. The multimedia version also puts what Deleuze and Guattari call 'planes of composition' together (in this case, the sequences in Jitter) only to disrupt them, making a non-linear text considerably less linear. Here non-linearity seems to be identified with increased intensity. Massumi (2002:26) suggests that 'intensity would seem to be associated with nonlinear processes: it creates resonation and feedback that momentarily suspend the linear progress of the narrative present from past to future'.

In addition, the multimedia version of the piece detaches emotions and perceptions from a sustained monolithic point of view, creating affects and percepts. This happens because the multimedia piece breaks down—much further than the print version—the sense of specific states of affect and perception rooted in particular subjects, though the process is already beginning in the print version. These affects and percepts are not distinct states, linked by causes and effects, but are multiple, simultaneous and superimposed. So the multimedia version puts the focus—even more than the print version—on the relationality between texts and the relation between text, image and sound. The ascending and overlapping cycles of sound, described above, also communicate a sense of changing and simultaneous intensities rather than evoking a particular emotional state.

We can also see here that this version produces its ethical and political content through affective intensities rather than sustained representation or exposition of political issues. The screener/reader catches glimpses of political and ethical meanings rather than detailed political insights and these meanings are extended, fractured and transmogrified by the addition of the sound and the transmutation of text into image. This is, of course, in keeping with objectives of avant-garde art, which has always conveyed political meanings in ways that are fragmented, anti-representational and inter-media. Relevant also here is the work of Tim Woods and Andrew Gibson, who theorise outwards from the theoretical stance of Emmanuel Levinas to argue that an ethical writing does not have to be determinate or representational. This is because ethics are not built on a foundational or fixed morality: for Gibson (1999:16), it is not based on 'categories, principles or codes' and does not presume 'an exteriority comprehensible in terms of hypostasized essences, static identities or wholes'. For Woods, similarly, an ethical poetry does not involve totalities and totalising structures, but emerges in fragments, gaps and fissures typical of an alternative tradition of poetic writing from Gertrude Stein to American language poetry. An ethical poetry, according to Woods (2002:255), resides in a 'poetics of interruption'. This poetics of interruption is also a vehicle for affect: in 'soundAFFECTs', the political and ethical become affective through the increased intensity brought by the visual and sonic, despite the gaps in the meaning.

On the other hand, the work of Oatley, with some qualifications, can also shed light on this transition from print to multimedia. To adopt Oatley's theoretical stance is to view the work from a more cognitive and empirically based position and any extrapolation we make from his ideas could be tested only in an experimental/empirical context. In order to do this, we have to adapt Oatley's ideas to the reading/reception process, which he does not himself do. In so doing, we can speculate that the experience of viewing 'soundAFFECTs' is that of forming a plan made up of sub-plans, which need to be coordinated. Throughout the work, we can hypothesise that there is a push and pull for the reader/screener between the fulfilment of the overall plan (the absorption of the multi-sensory work) and the interruption of the plan by the sub-plans (the reading of the texts). This interruption gives rise to a state of rapidly fluctuating arousal.

The main problem with Oatley's approach, from a cultural theory perspective, is that it retreats into a more subject-based, humanist position and that it falls back into the idea of emotion. Oatley's theoretical framework, however, has the advantage of being very concrete and precise. It could be used as the basis for empirical research to explore matters of reception and how reactions of the audience differ when reading the text, viewing the multimedia work or experiencing a combination of the two.

To conclude, the translation of the print version 'AFFECTions: friendship, community, bodies' into 'soundAFFECTs', the multimedia version, provides a changed affective experience characterised by a rapid flux of sensations rather than sustained emotional build-ups and identifications. To account for this transition, and the impact of the multimedia work, we require a theoretical framework that engages with the idea of affective intensities put forward by Deleuze and Guattari. It is also useful, however, to take on board Oatley's concept of emotional interruption and its possible applications. Ideally, an appreciation of the piece would involve reading 'AFFECTions' and viewing 'soundAFFECTs', since both versions evoke different types of meaning and affective response, which can be seen to be mutually enriching; here translation becomes a two-way, symbiotic process. Reading texts, listening to sound or looking at images is increasingly part of a multi-modal experience that requires negotiation between different media, technological environments and affective experiences. Such a multi-modality points to a huge diversity of textual possibilities, an increased range of aesthetic and cultural modes of production and to translation as a continuing and continuously evolving process.

References

Benjamin, W. 1999, *Illuminations*, Pimlico, London.

Brewster, A. 1996, 'Fictocriticism: undisciplined writing', *First Conference of the Association of University Writing Programs*, pp. 29–32.

Brewster, A. and Smith, H. 2003, 'AFFECTions: friendship, community, bodies', *Text*, vol. 7, p. 2, <http://www.gu.edu.u/school/art/text/oct03/brewstersmith.htm>

Dean, R., Whitelaw, M., Smith, H. and Worrall, D. 2006, 'The mirage of real-time algorithmic synaesthesia: some compositional mechanisms and research agendas in computer music and sonification', *Contemporary Music Review*, vol. 25, p. 4.

Dean, R. T., Brewster, A. and Smith, H. 2004, 'soundAFFECTs', *Text*, vol. 8, p. 2, <http://www.gu.edu.au/school/art/text/oct04/content.htm> or <http://www.gu.edu.au/school/art/text/oct04/smith2.htm>

Dean, R. T., Brewster, A. and Smith, H. 2008, *The Erotics of Geography: Poetry performance texts, new media works*, Tinfish Press, Hawai'i.

Deleuze, G. and Guattari, F. 1987, *A Thousand Plateaus: Capitalism and schizophrenia*, University of Minnesota Press, Minneapolis.

Deleuze, G. and Guattari, F. 1994, 'Percept, affect, and concept', *What is Philosophy?*, Verso, London and New York, pp. 163–200.

Gibson, A. 1999, *Postmodernity, Ethics and the Novel: From Leavis to Levinas*, Routledge, London and New York.

Hayles, N. K. 1999, *How We Became Posthuman*, University of Chicago Press, Chicago.

Kerr, H. and Nettelbeck, A. (eds) 1998, *The Space Between: Australian women writing fictocriticism*, University of Western Australia Press, Perth.

Kramer, L. 2002, *Musical Meaning: Toward a critical history*, University of California Press, Berkeley and Los Angeles.

Manovitch, L. 2001, *The Language of New Media*, MIT Press, Cambridge, Mass.

Massumi, B. 2002, *Parables for the Virtual: Movement, affect, sensation*, Duke University Press, Durham and London.

Meyer, L. B. 1956, *Emotion and Meaning in Music*, Chicago University Press, Chicago.

Oatley, K. 1992, *Best Laid Schemes: The psychology of emotions*, Cambridge University Press, Cambridge.

Oatley, K. and Jenkins, Jennifer M. 1996, *Understanding Emotions*, Blackwell, Oxford.

Smith, H. and Dean, R. T. 1997, *Improvisation, Hypermedia and the Arts Since 1945*, Harwood Academic, London.

Terada, R. 2001, *Feeling in Theory: Emotion after the 'death of the subject'*, Harvard University Press, Cambridge, Mass., and London.

Woods, T. 2002, *The Poetics of the Limit: Ethics and politics in modern and contemporary American poetry*, Palgrave Macmillan, Basingstoke, Hampshire.

Endnotes

[1] A version of this chapter was previously published in *Scan: Journal of Media Arts Culture*, vol. 4, no.1, 2007.

[2] See CD-ROM accompanying Dean et al. (2008) for a full-screen version of 'soundAFFECTS'.

[3] There is, however, also a web version of the piece available at <http://www.textjournal.com.au/oct04/smith2.mov> (Dean et al. 2004). This is a Quicktime movie that 'captures' one rendering of the piece.

[4] As Lev Manovitch (2001:47) says, 'In new media lingo, to "transcode" something is to translate it into another format.'

[5] Massumi also draws a distinction between emotion and affect. For him, emotion is conventionally and culturally constructed and affect is intensity, which is 'unassimilable' (Massumi 2002:27–8).

[6] The relationship between new media writing and affect has also not yet been adequately theorised. This is partly because of a paucity of theorising about new media writing in general and partly because of an understandable tendency in early theorising to focus on technical features such as hyperlinking and interactivity rather than their psychological and political effects.

[7] John Howard was Australian Prime Minister from 1996 to 2008.

[8] UNSW here stands for University of New South Wales.

[9] Another section of the collaboration, 'Frrrustration', is also a good example of affective intensities, in a slightly different way, because it is composed almost entirely of bodily sensations conveyed in a build-up of short phrases: 'the body takes over like a crazed mechanical wind-up toy, rehearsing the pretext of metaphor. its elegiac pulse thumping like a jackhammer, its discursive fluids leaking, bones grinding, muscles clenched in this Olympian task of speaking.'

[10] Technical details of the programming and performance of the image and sound were provided for me by Roger Dean.

2. *Edible Audience*: what about this gastronomic performance translated as sound art?

Alistair Riddell

Introduction

If new media performance continues to evolve through the convergence of diverse technologies, where does this leave the audience with respect to the creative experience? Is it important for the artist to understand to what extent and how the audience understands the use of technology in their work and how that technology facilitates conveyance of the concept? Should the artist care? This chapter discusses the relationship between artist and audience through an examination of a complex sound-art project and seeks to initiate thoughts from which artists might contemplate the form of future projects and their subsequent performances.

The intention of the *Edible Audience* project[1] was to articulate a concept through the use of an augmented-reality system, which integrated images into live video projection and controlled sound diffusion, all through performer interaction. There was a noticeable absence of the technology during the performance, with the emphasis being on the performers, the sound and the projection. This will be discussed in detail later in the chapter.

The concept was really quite simple and straightforward with a humorous, if rather dark perspective on the nature of consumption in contemporary society. The performance action centred on the 'consumption' of images of the audience and body parts and was formally structured as a continuous narrative around the courses of a meal: entrée, main course, dessert and a toast. Although the technical implementation and the performance mostly ran smoothly,[2] the impression from discussions with a few witnesses raised the question of how unified, clear and effective the event was as sound or data art.[3] It was not that the range of comments was so diverse but that few, if any, seemed to concur with concepts and impressions that we, as performers, held as somehow axiomatic or fundamental, and consequently comment worthy.

Almost immediately after the event and a gestation period of reflection, the question arose of whether this was a fundamental point of concern for sound/data media artists in general. Many of these works tend towards an expression of concepts that require, on the part of the audience, critical and reflective interpretation during and after the event. Undertaking this contemplative process is now more acute in this age of vigorous technical experimentation. Before discussing this in detail, however, let us slip back in time and consider a sound work that contributes to a trajectory of contemporary art evolution under which the *Edible Audience* project squarely lies.

From the past

More than 50 years ago, John Cage composed *Imaginary Landscape No. 4* (Pritchett 1993:89–90; Revill 1992:143–4). That it was 'composed' set the work in a particular musical context even though the work itself was anything but conventional. The work was scored for 12 radios and two performers were required to change the frequency, volume and timbre settings on each radio while distributed around the performance space.

It is my view that Cage's precise instructions to the performers establish, in the minds of the audience, a clearly delineated work with a predetermined aesthetic objective and form. That the performers are not necessarily highly skilled at manipulating the radios' controls is not seen as an impediment to the outcome of the work. In part, the performance relies on the audience accepting a fundamental degree of skill and the ability of the performers to follow the 'score'. The decoding of this performance information is what sets the performance apart from some kind of random activity. It establishes performer competency with a sound source that is readily understandable.

An important distinction between the 'radio' and a traditional musical instrument can be understood from a statement by Fels et al. (2002:110): '[o]ne of the key attributes of instruments required for adoption into the literature is expressivity; this is a necessary condition for acceptance'. It is easy to appreciate that the 'radio' is not an expressive instrument in the traditional sense but it could be argued that the expressiveness of the radio in a performance such as *Imaginary Landscape No. 4*. lies in its ability to effectively articulate the concept. It is clear that in this case a traditional instrument would not be appropriate.

There are a number of Cage's works that easily raise the question of what musical skills are really required for the performance; *4′ 33″* is probably the most widely known example. Such reflection is, however, largely neutralised—at least in retrospect—by the objective of conveying the concept and the exploratory aspirations of the composer. As a consequence of Cage's works and influence, the remainder of the twentieth century accumulated a vast repertoire in which the question of musical performance virtuosity could be legitimately questioned if it were not for the fact that there was little precedent and the exploratory agenda was in full flight.

Although *Imaginary Landscape No. 4* was conceived as a musical work, it was part of the genesis of sound-art performance practice in which a technological presence was essential to the expression of the concept.

Closer to now

Exploring and experiencing radically new concepts can be seen as the principle objectives of the contemporary artist and something of an expectation for the audience from the mid twentieth century onwards. This, of course, is an agenda that varies. Often in historical examples, the performance practice was transparent, as in *Imaginary Landscape No. 4*, and in itself did not cause the audience to wonder too much about how the sound was produced and controlled. The gradual increase in the use of and dependency on technology in performance, however, began to change that. Technology added new layers of abstraction and obfuscation unprecedented in public performance. The audience either had to accept what the technology did or ask questions about how the technology was involved. This dilemma persisted and, over time, it became apparent that works using new technologies were less well received by the audience as the novelty and euphoria of the 'technology revolution' wore off. Something of a catch-22 existed here because many of these performances were predicated on an overt use of new technology. Rendering this issue moot, however, could be achieved in a variety of ways but essentially depended on taking the audience's mind off the question of what the technology was doing and focusing them on the artistic concept.

By the end of the twentieth century, context and spectacle were playing increasingly important roles in the presentation of work that involved technology. It was more and more apparent that if one used technology in a manner that attempted to showcase it or its function the audience was inclined to consider this less artistically significant. The better approach was to integrate the technology into a more complex event. Not all integrations, however, are successfully implemented and the configuration of such events remains an elusive undertaking. Manovich gives some indication of this challenge while attending an event in St Petersburg in 1995:

> Under the black hemispherical ceiling with mandatory models of planets and stars, a young artist methodically paints an abstract painting. Probably trained in the same classical style as I had been, he is no Pollock; cautiously and systematically, he makes careful brushstrokes on the canvas in front of him. On his hand he wears a Nintendo Dataglove, which in 1995 is a common media object in the West but a rare sight in St. Petersburg. The Dataglove transmits the movements of his hand to a small electronic synthesizer, assembled in the laboratory of some Moscow institute. The music from the synthesizer serves as an accompaniment to two dancers, a male and a female. Dressed in Isadora Duncan-like

> clothing, they improvise a 'modern dance' in front of an older and, apparently, completely puzzled audience. Classical art, abstraction, and a Nintendo Dataglove; electronic music and early twentieth-century modernism; discussions of virtual reality (VR) in the planetarium of a classical city that, like Venice, is obsessed with its past—what for me, coming from the West, are incompatible historical and conceptual layers are composited together, with the Nintendo Dataglove being just one layer in the mix. (Manovich 2001a:5)

Manovich acknowledges and attempts to frame, in a positive and compelling way, an event that clearly challenges the audience. Couched in personal nostalgia and historical references, such a description suggests the kind of cognitive engagement needed for events that drive the senses from multiple directions with occasional tenuous connections.

While it might be the case that some performances—and these could be regarded as successful in one respect—appear to the audience as cohesive and integrated in the presentation of the artistic statement, other more experimental works are problematic in this respect. Assuming that the performance unfolds as planned, an initial starting position for gauging external success might consider the points of mutual understanding and the relationship between the performer's and the audience's reading of the event. It, however, now clearly extends beyond the presence of performers. Pedro Rebelo offers a way to think about the nature of the space inhabited by the performers and the audience:

> The notion of performance itself implies a somewhat nonlinear environment. While the performer has some level of control over the environment and potentially over the performance instrument, it is the uncontrolled, the chance events, the risk, that defines the performance environment. It is the 'nonlinearity' that is responsible for the chemical reaction between a performer and her audience. (Rebelo 2003:182)

Interpretation from a basic cognitive state

Given that the performance context can be 'non-linear'—that is, possessing changing and evolving connections between the parts of the performance—it is important to establish a basic set of criteria from which the audience might initially perceive, question and evaluate the broader structural and configuration elements. Ideally, an audience has a basic level of pragmatic experience from which to assess and understand a concept's implementation objectively. These experiences would operate subconsciously and go to moderating preconceptions and highly subjective impressions. This might, however, be a conservative and anachronistic view of contemporary audiences.[4] Given that such a starting point should be more inclusive and should acknowledge the complexity of the audience's background, here is a simplified mapping of basic referential experiences, which loosely fall under four categories:

1. recognition of expert, skilled or authoritative performers
2. acknowledgment of the function and role of technology
3. understanding of a shifting performance emphasis to spectacle and context
4. ability to perceive combinations of the above categories.

1. Underlying most people's experiences and expectations of sound art are performance expertise, skill and aesthetic authority. All forms of instrumental and vocal music demand that the performer aspires to or has achieved a level of competency that allows them to convey the essential point of a musical work.

In more contemporary forms of sound performance, such as those using electronic and computer-based technology and DJs, perception of skill and expertise is more difficult to assess due to a lack of the codification of the practice and the fact that the dynamics of the art are based on changes in technology and the momentary aesthetics of sound. Even though the DJ might have acquired a high degree of technical facility, the evolution of the music is predicated on learning how to put sounds together to create new genres. The audience, however, knows precisely when a contemporary artist has been successful in performance. In the case of the DJ, there is a collective energy among the audience that identifies success. This 'social factor', as discussed by Winkler (2000:2), is in this instance at its most empowered and expressive, often overriding personal inhibitions and any sense of intimidation among individual audience members.

2. The use of technology therefore does not necessarily diminish competency but changes the relation between the resulting sound and the performer. The use of mixers and sound-diffusion hardware comes at the end of a chain of sound production. Technology can increase the complexity and sophistication of the performance without continuous performer interaction. The performer can concentrate on higher-level aspects of the sound production and performance structure.

3. Manovich's observations in the earlier quote suggest that contemporary performances involving technology are likely to embrace spectacle and context. Whether this is a result of the dilemma between an increased use of technology and a general sense that the audience doesn't want to be a guinea pig for new technology experiments is hard to clarify. Certainly, it is likely that there is a general weariness towards technology brought on by decades of hype and promise, promulgated through the mainstream media. Technology, however, remains a creative incentive but now modulated by a need to direct its potential towards the lucid articulation of creative concepts.

4. The evolution of sound-art practice draws together all of the above points but perhaps is dependent less on the refinement of technical skill and more on

the rapid uptake of new technologies and a necessary grasp on its creative potential. There is no universally recognised standard of skill. Most sound-art performances continue to depend on innovative use of technology and highly creative outcomes.

The question of whether the audience's understanding and experiences should be reduced to such basic criteria inevitably arises and is open to criticism. The intention here, however, is that it initiates an objective starting point for the artist to consider the audience as a whole before the manifestation of the work. In this respect, the artist might ask the question of whether the work meets all or some of these points and even whether there are others that might be crucial to appreciating the performance.

The audience perspective

Performances involving mixed media, interactivity and collaboration, especially those of a distinctly experimental or avant-garde nature, are difficult to articulate or critique for most audiences even though what constitutes the 'audience' today is more ambiguous than in earlier times. One might be advised to consider the entire audience as a collection of artists in their own right, some simply less familiar with the genre or concept.

The *Edible Audience* performance raised the central question of how and to what extent the audience experienced the implementation and interpreted the thematic elements (theatre, image, sound and interactivity). It was assumed, on the basis of the programmatic thread of the 'dining' performers, that the audience would link the event action. It was, in a sense, a narrative structure but the performance resembled that of the early 'silent movies' in which the performers were playing/controlling the sound while providing the source action for the video projection. In reality, the performance probably looked nothing like our collective idea of a 'silent movie' but the analogy facilitates a way of thinking through the network of connections between action, sound and image.

Was the audience able to follow the structure as a singular, brief and complex presentation as theatre, as sound and as projected image? Was the convergence of these parts cohesive enough for the audience to take in the central conceptual point: the issue of consumption in contemporary society? Did we as performers strive to make the concept clear? Given that Liquid Architecture 6 was billed as a sound-art event, was the audience predominantly distracted or confused by the amount of audio/visual information? Did the program promote to the audience and support in practice the ideal of experimental works?

In seeking answers to these and other questions, or an understanding of the performance in retrospect, it is worth considering how the project space impacts on the experience. Rebelo provides a general starting point:

A performance space, in the context of nonlinear digital media structures, implies sophisticated analysis in the areas of gesture, one-to-many communication schemes, individual presence, idiosyncratic action, and instrumentality. The performing body operates in a space of expectation, in a space that tends 'towards more or less coherent systems of non-verbal symbols and signs'. (Lefebvre, quoted in Rebelo 2003:182)

The challenge in considering Rebelo's text lies in how to identify and assess the levels of communication, and by whom and when. Some modalities of communication can be considered during rehearsal but optimising these might not make a significant difference under performance conditions. Impressions of success and failure depend on individuals who might or might not be adequately informed about the work. The *Edible Audience* project itself was part of a larger program and the diversity of works presented, contrasted, complemented and possibly clashed, thus making an intelligent, impromptu and comprehensive critique of any one work difficult without substantial qualification and reflection.

Those in the audience familiar with the constituent components of sound-art performance practice were possibly able to navigate the various levels of activity and thus extract from the experience a more positive memory. It is accepted that such a performance, operating under conditions of, at times, randomness and explicit but arbitrary control, will have moments in which the aesthetic flounders, only to be asserted at a later time. This instability, this ebb and flow of cohesiveness, lies at the heart of all improvised performance in which generation and control of events become a moment-by-moment concern for the performers.

Sound-art practice is, by definition, innovative, exploratory and woven into a configuration mapping action data from diverse media systems to a concept. While the data might not be subject to direct human intervention, today it is more common that there is some kind of interactivity or influence over the data that controls, generates or constitutes visual or sonic manifestations. The following comment could apply equally to complex forms of sound-art presentation:

Interactivity may offer an entirely new approach to music-making, and so in order to avoid getting stuck in the current musical paradigms, we should question not only the nature of the system input (such as musical notes, tempi, rhythms, or human gestures, dance movement, or conductor's gestures), but we should pay equal attention to the output of the system, and the qualitative relationship between the two. (Paine 2002:297)

It might be assumed that the question of audience comprehension lies in how they perceive the mappings in terms of action or object to aesthetic results. Can we assume that the more abstract or esoteric the artistic concept, the more the audience will struggle to engage with the work? No. It depends on the mode of concept manifestation and whether the audience has been briefed on what to expect through other means before the performance.

Categorical overview of *Edible Audience*

Few artists have the opportunity to question each audience member about a performance in a formalised manner. Typically, incidental conversations, impromptu and casual responses vary greatly and often focus on matters that the artist might consider as circumscribing their particular concerns. The audience experiences only a brief moment in the life of the project. The artist, however, will be aware of a general 'feeling' about the success of the performance, usually at its most poignant between the conclusion and the applause.

Irrespective of the outcome of the performance, it is valuable for the artist to review the concept, implementation and performance, not only from their own perspective but what they think it might have been like for the audience. Consideration of a sound-art performance from the audience's point of view, however, invariably leads the artist back to an examination of the entire event.

Before undertaking a more detailed account of the project, it is worthwhile reviewing key aspects of the experience under a number of critical categories.

Concept

- A critique of our consumer-driven society
- the theatre of four diners at a table eating images of people
- use of an augmented-reality system for live performance
- engagement with a kind of sound-art theatre.

Technology

- The use of two networked computers to implement vision, image and sound processing
- multi-channel sound
- deployment of the technology in the performance space
- configuration of the performance props and the camera
- configuration of AVIARy, the augmented-reality software application
- development of SuperCollider 3 audio patch.

Motivation

- Distribution of expertise
- learning experience
- uniqueness of the event
- institutional collaboration.

Agendas

- Performing with AVIARy software
- establishment of performance profile through a new sound-art project.

Performance

- No practical familiarity with the performance space
- lengthy set up
- multiple performers
- restricted but vague performance configuration
- few rehearsals
- no substantial performance training or experience with such interactive systems before the project
- limited understanding of the totality of the project presentation from an audience's perspective
- inability to hear or see what was intended for the audience
- effectiveness of the performance interface.

Context/venue

- Acoustic properties of the space
- image presentation
- stage area
- sound diffusion set-up
- program structure.

Audience

- Comprising those who are:
  —practitioners of sound art
  —familiar with sound art
  —unfamiliar with but interested in sound art
- their seating location in the venue space
- those able to maintain interest through a multi-layered-sound art presentation.

The ordering of these categories might be seen in retrospect as important from an analytical point of view but if considered before the performance they might need to be constantly reordered to help predict and control the outcome.

Performance anatomy

Although as a performer I was not directly part of the audience, an analysis of the structure of *Edible Audience* should provide a way to examine what was presented to the audience and reflect on what was important and was not, from at least one perspective.

The controlling software application was AVIARy,[5] an 'augmented-reality' system based on the AR toolkit (<http://www.hitl.washington.edu/artoolkit/>). AVIARy analysed live video input and searched for predefined patterns, called 'fiducial markers'. When these patterns were recognised, other images were superimposed on the markers, thus hiding them. These images could be moved around with the patterns and had six degrees of movement: X, Y, Z (height), yaw, pitch and roll.

On recognising a fiducial marker, AVIARy would also output OSC datagrams over a network connection to the audio application, SuperCollider 3, running on another computer. The purpose of this was to reduce the computational load on the machine running AVIARy. The sound—processed audio files and some real-time synthesis—was diffused through an eight-channel sound system, which included two sub-woofers.

There were six performers: four diners and two waiters. Although there were a number of rehearsals and trials with various technical and performance configurations, it was clear that improvements in the performers' roles would be possible only with more performances and rehearsals.

The fiducial markers were handled by the performers who were 'acting' out a dining scene, which as mentioned above, was broken up into several acts (entrée, main, dessert courses and a toast). Interludes indicated to the audience what was going to happen next.

The performers were to the audience's left, with the projection screen centre stage. Perhaps they should have been located in front of the screen to enhance the idea of making a connection between the theatrical action and the projected image but that is only speculation. It might have made the experience more confusing or it might have had no significant effect at all. The audience saw mostly the site of the performance with not much detail. The acting deliberately exaggerated gestures but the fiducial markers were small and probably not recognisable from a distance. They were, however, occasionally visible on the main screen. This should have given the audience a clue as to the relation between the markers and the superimposed images.

The performance started with a title screen triggered by a particular marker on the table. This marker was then replaced with an 'entrée' title marker. The first diner entered and sat at the table. A plate of four entrée markers was served and the diner began to move one of four unique markers around. At this point, the audience could easily observe the relationship between the marker, the image on screen and the movement of the sound.

One by one, the other diners entered and joined in the performance, selecting markers and moving them around in the camera space. It can be assumed that not much of the theatricality of the performance at the dining table was visible to all the audience. What the audience could see clearly was the video projection of the table top, where the markers and the performers' hands could be seen with the addition of the images superimposed by AVIARy (see <http://www.petermorse.com.au/old/PMcomhtml/petermorse.html>).

From a visual perspective, then, two points of reference existed but were perhaps not easily correlated due to size, position and perspective. There was also a small delay between the movement of the performers and the image projected on the screen but because both could not be viewed at once, the delay was probably not a significant factor in the visual experience.

The next layer of experience existed between the movement visible in the projection and the sound. Even though the relationship between the performer's movement of the markers and the sound was in real time, to make this explicit, the performers had to know how to optimise their movements to make the spatial movement of the sound very defined. The movement of the performers' markers across and around the table corresponded with the movement of the sound in the audience space. This would have been very clear if there had been only one performer with one sound but there were four and the sounds were not always that dissimilar or always in discrete locations.

In addition, the visual mapping on the screen did not immediately correlate with the audience space. If the audience figured out how to view the screen as compass bearings, they might have been able to appreciate how the sound was mapped into their space. This might have been understood through listening and making connections with certain performers' movements, but could have been difficult and elusive with four active performers and rapid changes in sound.

The question of whether the sound/movement correlation was important for the audience to experience and whether it added anything to their experience would probably be answered in the affirmative from the artist's perspective. Appreciating the mapping strategy (what the performers were doing and what happened visually and sonically—and there clearly was some challenge here) might, however, be argued as simply an intellectual exercise, not necessarily crucial to a full appreciation of the experience. All that might have been required on the part of the audience—thus avoiding the tedium of technical detail—was that they understood that there was a correlation.

The other issue was that the performers themselves were not experienced in moving the markers with a specific understanding of how the sound moved. Even learning how to do that was problematic given the nature of the interface. Talking about the 'Iamascope', an interactive kaleidoscope that uses computer video and graphics technology, Fels et al. (2002:115) state:

> The player's movements are unconstrained and the player has to discover the mapping on his own. The closest metaphor is that the interface is

> like a ten-string guitar where the computer holds down the chords automatically. The player strums the strings by moving in the bins. While this metaphor helps make the mapping easier to understand it does not help in learning to play the device. This is because the metaphor is not quite accurate.

And later:

> In general, this attribute of free hand or free form gesture mapped to sound is problematic. Very few metaphors provide a strong enough link between gesture and output to provide an easy-to-learn mapping. Thus, even if the metaphor and the mapping are easy to understand, they will not necessarily lead to a very expressive instrument. In this situation, other paths to achieve transparency need to come into play to make the instrument expressive. (Fels et al. 2002:116)

The task of mapping gesture to action for the *Edible Audience* performers was made even more difficult because while it was important to try to keep the markers visible to the augmented-reality system, it had to be done without looking at the projection screen. This would have detracted from the theatrical nature of the performance. If the markers were not identified by the system, the superimposed image and the sound were also not present.

Another problem was that the performers were not in the same specific audio space as the audience, so the performers could never really hear the sound moving in response to their hand/arm movements. It was possible to hear changes in some locations of the sound if the performer was concentrating on their sound while moving a marker.

What, of course, would have improved the performance immensely would have been more rehearsals. It is probably symptomatic of many sound-art performances that the performers suffer from a lack of experience, confidence and certainty, which must be conveyed to the audience at the beginning. If the performers know that what they want to convey can be communicated, that expectation must pervade each subsequent performance.

Surveys

Finally, it is not the intention here to critically consider the field of arts surveys, for this most likely substantial research area is simply beyond the scope and purpose of this chapter and is best left to those who understand the specific processes, techniques and outcomes.

The question, however, had clearly been lingering and waiting to be asked: 'Why not conduct a survey, either entry, exit or both?' The logic of this is unmistakably compelling. Surveys are used extensively in our everyday world and, properly conducted, can result in valuable information. Such information, if properly gathered, would have, at least, the appearance of being frank, objective and reflective within the recent period of experience. In the case of the *Edible Audience* project, a survey would have been an additional burden that none of us would have been able to undertake.

It would have been difficult to structure in light of the fact that we were constantly reworking the performance practice and configuration. The performance was short and likely to happen only once, so we focused on that.

As the *Edible Audience* performance did not go entirely as planned or anticipated and we felt the denouement was somewhat compromised, we as performers might have considered a predefined survey as somewhat invalid, assuming that we had prepared questions on the basis of a flawless performance. The logical thing to do, therefore, in the preparation of questions, would be to accept that the performance might not go as planned. If that is the case, what questions could we ask that we would regard as important to future projects? While it is likely that there are such questions, it would take time and experience to fully appreciate what they are.

If there had been the opportunity to perform the work again with the technical problems overcome, a survey would be more valuable at that time rather than previously. This assumes that what the project is conveying is now more clearly evident. Should surveys be conducted only at those times?

At this point, it is clear that the question of a formal survey for a singular event within a larger experimental performance context can probably be dismissed as unnecessary, distracting and burdensome. A survey conducted independently and inclusive of the larger event is another matter and probably would have a different agenda.

Possibly, of more importance and value would be for the performers/artists to know what questions to ask at those impromptu moments when talking with people who attended the performance. This usually happens some days or weeks later. If, after the performance, the artists were to discuss the performance and collectively arrive at questions to ask at those appropriate moments, this would be the most useful form of information gathering given the nature of the event. An issue that arises from this is that some artists are not always able to accept, interpret or respond appropriately to criticism. They simply lack the objectivity and detachment necessary to appreciate views from a range of audience members, rather than from a select group of friends whose critique might be personally biased or influenced by a priori knowledge.

Conclusion

This chapter, then, constitutes a singular and private attempt to raise one artist's perspective on the reception of a sound-art performance. It should be viewed as the beginning of a process of self-evaluation of practice with the objective of strengthening the relationship between concept, performance and the audience.

It is expected that under the vagaries of performance, any implementation might be compromised before and during the event but not after. Conveying publicly that the project has been broadly thought through rather than concerned primarily with the performance mechanics should compensate for performance irregularities and hopefully the audience will recognise these for what they are and still appreciate the overall intention in the work.

In the case of *Edible Audience*, was the audience able to navigate and connect the performance as theatre, as sound and as projected modified video? The answer to this is probably a cautious 'yes'. The audience had to at some point make a connection between the performers' actions and the video image then make frequent updates during the course of the event. The question of the perception of the performers' relationship with the sound is more complex because the mapping of the sound to the action was less evident, especially for that part of the audience who could not clearly see the performers. The sound appeared to be understood as somehow separate or with a more obscure connection from later discussions. Even the performers struggled with this aspect of the performance. In addition, from discussions arising after the event, it was clear that some people concentrated on singular aspects of the performance depending on their cultural interests—for example, the visual display or the sound. These were compelling and more easily focused on than the theatrical component for which one had to be closer to the front to see clearly. This raises the question of how perceptually unified the performance was but also suggests that it was flexible enough for the audience to maintain interest and shift their attention across the detail of the performance.

The *Edible Audience* critique of 'consumption in contemporary society' was probably not articulated effectively enough to the audience before the performance but whether this really mattered is unclear. In later general discussions about the performance, it was clear, however, that this could be understood as a subtext or a form of explanation and justification of the performance on the part of the performers/artists. From the audience's perspective, this critique, if it was understood at all, seems to have been viewed independently of their experience. One reason for this is perhaps that the performance did not explicitly or controversially take a sociopolitical stance other than to present, in a surreal context, the idea of consuming fellow humans as a logical consequence of our consumer society.

The achievement of presenting such an experimental project went a long way to creating a sense of success even though there was one glaring problem during the performance and a mild sense of disappointment that real food could not be used. On the other hand, the audience largely sensed that the performance was cohesive and performed without obvious or disruptive technical problems.

Thinking of how the audience might receive a performance does not mean pandering to them by making the work transparent to the point of shallowness but rather is a way of ensuring that the optimal conditions for concept and performance presentation are achieved. This is by no means easy to understand and even less so if it has not been thought about at all. In developing the *Edible Audience* project, we did from the outset think about the audience. We tried to include them and directly inspire a critical perspective on consumerism, as we are all part of our immediate community and society. That could have been confronting if understood and contemplated throughout the performance but nevertheless was a significant achievement for the project. I personally felt that this was not clearly understood. The theatre and media presentation formats were complex and required constant engagement by the audience. In effect, the performance was too demanding and the concept could be considered only in retrospect.

From the analysis of one sound-art work, it is not possible to generalise exhaustively on the effectiveness or shortcomings of all future performances but there are questions in many forms that can be asked and addressed on the basis of any creative experience when considering undertaking another. The point of formulating and responding to such questions is to accumulate, over time, a body of experience and mode of thinking about a project before any practical undertaking. While the final outcome cannot be fully predicted, understanding and analysing aspects of a project in the very early stages can indicate and highlight issues that might, if acted on, make a significant positive difference.

Acknowledgments

I am indebted to Professor Stephen Barrass for the discussion that sparked the subject of this chapter. Also, I wish to thank my fellow 'consumers' who performed in the Liquid Architecture 6 Sound Art Festival at the National Gallery of Australia on 23 July 2005: Stephen Barrass, Tim Barrass, Anita Fitton, Peter Morse and Onaclov.

Bibliography

Fels, Sydney, Gadd, Ashley and Mulder, Axel 2002, 'Mapping transparency through metaphor: towards more expressive musical instruments', *Organized Sound*, vol. 7, no. 2, pp. 109–26.

Kostelanetz, Richard (ed.) 1971, *John Cage*, The Penguin Press, London.

Manovich, Lev 2001a, *The Language of New Media*, MIT Press, Cambridge, Mass.

Manovich, Lev 2001b, *The Anti-Sublime Ideal in Sound Art,* viewed 5 September 2005, <http://www.manovich.net/DOCS/data_art.doc>

Paine, Garth 2002, 'Interactivity, where to from here?', *Organized Sound*, vol. 7, no. 3, pp. 295–304.

Pritchett, James 1993, *The Music of John Cage*, Cambridge University Press, Cambridge.

Rebelo, Pedro 2003, 'Performing space', *Organized Sound*, vol. 8, no. 2, pp. 181–6.

Revill, David 1992, *The Roaring Silence*, Arcade, New York.

Winkler, Todd 2000, 'Audience participation and response in movement-sensing installations', *ISEA*, Paris.

Endnotes

[1] *Edible Audience* was performed as part of the Liquid Architecture 6 Sound Art Festival at the National Gallery of Australia in July 2005 (<http://www.nga.gov.au/LiquidArchitecture/bios.cfm>). I had forgotten until reminded by Alistair Noble at the time of editing this chapter that *Edible Audience* was similar in concept to a performance presented by Cage and other artists in New York as part of the New York State Council on the Arts Contemporary Voices in the Arts series of 1967. He said, 'On stage, a five-course dinner was served, during which the performers' voices and sounds of the silverware were electronically amplified' (in Kostelanetz 1971:55-56).

[2] There was an unexpected and not hitherto experienced sustained false triggering on a pattern that caused images and sound to appear sporadically throughout the performance. They should have appeared only at the end.

[3] I prefer the term 'sound art' as defined by Manovich to 'new media', the former term being more relevant where diverse digital technologies are combined in one performance. Although a 'sound-art' festival in this case, the term emphasises what was important in this project.

[4] See a later discussion of this in the text.

[5] AVIARy was written by Tim Barrass.

3. Translating the musical image: case studies of expert musicians

Freya Bailes

There is more to music than the production and perception of real sounds; musical experience also involves musical thought through imagining and mentally re-presenting sounds. This can occur unintentionally, as in the phenomenon often called 'tune on the brain'. Alternatively, imaging music can be an involuntary corollary of musical activity, such as anticipating the next track on an album while listening to music or working towards an ideal musical sound in performance based on internally 'hearing' how it should sound. Finally, a mental image of music can be deliberate, as in the 'silent' analysis of a musical score or the *auralisation* of sound in harmony and counterpoint exercises.

How does musical thought in the form of musical imagery—a mental image of how the music should sound—translate to musical production? Musical imagery is the conscious experience of an internal representation of music, or *inner hearing*. Various definitions of imagery have been offered. In the auditory domain, Intons-Peterson (1992:46) describes it as 'the introspective persistence of an auditory experience, including one constructed from components drawn from long-term memory, in the absence of direct sensory instigation of that experience'. Musical imagery is a production of the *mind's ear*. It can differ importantly as an experience from the internal processes that channel and interpret any real incoming sensory information, otherwise known as *perception*. Imagination encompasses imagery, being 'the faculty or action of producing mental images of what is not present or in one's own experience' (*Collins New English Dictionary* 1997). The other definition of imagination (from the same source) as 'creative mental ability' presents imagery as a key ingredient in creative thought. Seashore (1938:161) makes a link between imagery and creation when he speculates that 'perhaps the most outstanding mark of the musical mind is auditory imagery, the capacity to hear music in recall, in creative work, and to supplement the actual physical sounds in musical hearing'. Robertson (1996:20) argues that 'we must accept that memory and a beautifully formed musical imagination are the prime factors in music making'.

Musical imagery has particular importance for musicians, functioning in creation, performance, auralisation, recollection and anticipation. Examples of its assumed occurrence include a performer developing an interpretation in her or his head, a conductor doing silent score reading in the absence of any instrument and a composer imaging new sounds or new combinations of sound. We know surprisingly little about the phenomenon of musical imagery, however, due in part to a reliance on introspection or inference based on a translation of the image into observable sound. What do expert musicians say about their imagery? For example, how important and how prevalent is musical imaging in expert musical activity, and how does it translate to and from perception? This chapter investigates these questions, with respect to three professional musicians in Western classical music: a composer, a pianist and a profoundly deaf organist. These musicians discuss the relationship of their mental representations of music to the music they perceive. First, an examination of existing research into the translation of image to sound, and sound to image, will be presented here.

Translating image to sound

The most obvious starting point when considering the translation of an auditory image to sound is the task of a composer. There is no shortage of anecdotal reports concerning the musical imagery of famous composers such as Beethoven, Stravinsky and Elgar. Mountain (2001:271–88) and Retra (1999) have attempted to undertake a more systematic understanding of the subject. For instance, Retra designed a study in which composers were asked to provide a verbal commentary on a composition task away from an instrument. The purpose was to investigate the nature of mental representation in the composition process.

Inspiration in composition is often taken to be synonymous with musical imagery and 'the myths that surround the one have confused investigation of the other' (Mountain 2001:273). One of these myths is that inspiration takes the form of a complete and pure auditory image to be translated in a sequential manner from the mind to paper. Mountain's evidence suggests that, in reality, composers are more likely to have been mentally working on music for a while, modifying and developing an image rather than transcribing one in virgin form. Harvey (1999) provides a comprehensive discussion on the subject of musical inspiration. Drawing on the reports of composers including Mahler, Dukas, Elgar, Boulez, Maxwell Davies, Schumann, Ligeti, Hindemith and Rubbra, he writes that 'the unconscious is clearly capable of reordering mental impressions to find solutions to compositional difficulties, without any need for conscious thought' (Harvey 1999:22). Penrose (1999) formulates a theory that creative thinkers unconsciously put up ideas for contemplation but consciously eliminate those that are redundant. Retra's (1999) conclusions are similar to this. She found that the composers in her study seemed to use imagery in compositional decision making as a means of holding information in conscious attention, inwardly 'hearing' it in what was metaphorically termed the 'mind's ear'. Mountain (2001:275–6) describes a similar intermediary role of mental imagery towards the final stages of musical composition: '[t]he vividness of the auditory image is…necessary…so that it can be clearly maintained and referred to during the sometimes tedious procedure of notation.'

The alternation between unconscious and conscious thought seems to be at the root of many misconceptions regarding the role of inspiration and the role of assimilation. A chronological dimension might mean that composing 'traces a path from the intangible imagination to the tangible reality of a created work' (Saxton 1998:6). Imagination is more than imaging, involving a degree of creativity over mere visualising or experiencing a ready-made copy. According to this, musical imagery might represent an intermediary point in translation between imagination and what Saxton (1998) describes as the 'aural detection' stages of creation. The process of developing a compositional idea implies a musical imagination, musical memory and the ability to alter and mentally rehearse an image.

Remarkably little research on musical imagery has looked at the translation of image to sound in performance, with the exception of Brodsky et al. (1999:370–92), who explored the inner hearing of orchestral musicians. Kvifte (2001:219–35) also explores imagery for performing musicians, describing the importance of mental imagery for the memory of form in Norwegian *Hardingfiddle* music. The author's main concern is to determine 'the relation between observed formal structure and possible internal images of them' (Kvifte 2001:219–20). In other words, the interest is in understanding what information about the performer's guiding mental image can be gained indirectly from their observable performance. This assumes a direct translation from mental imagery to overt behaviour, describing a feedback between one and the other in the process of performance.

In Western classical music, there is a tradition of translating visual and auditory imagery to sound when performing from memory. Marek reports that Toscanini was renowned for his musical memory and the veridical nature of his imagery:

> His memory was strengthened by what I may call the 'mind's ear', meaning the ability to hear a composition by reading it. That ability is essential to a conductor, but Toscanini possessed it to an amazing degree. He had but to glance at a page of complex music, his glance seemingly casual, and he heard the page both horizontally and vertically in his imagination. (Marek 1982:416)

Translating sound to image

Where does the musical image come from? A translation occurs from the perception of sound to the creation of a mental re-presentation of that sound, including the conscious auditory image. Musical composition is not purely endogenous, requiring as it does the initial perception of constituent sound material. Even in Western classical music, in which written notation might seem to be the principal carrier of musical information, 'Western art music is in fact dependent upon an "oral" transmission of performance tradition' (Godøy and

Jørgensen 2001:182). Great musical performers convey information additional to the musical score through the shared communication of a detailed musical image. Marek (1982:415) believes that 'retention of minutiae is an attribute of the interpretive artist; it lies at the base of performance, and it can be trained'. It is important to note the emphasis on the retention of perceived sound as a foundation for creative interpretation and the supposed potential to develop this skill.

Musical essence

'Take out the image from the musical mind and you take out its very essence' (Seashore 1938:6). Some musical traditions, however, are more dependent on auralisation than others. For example, consciously imaging sound away from a musical instrument (be that a violin, voice, CD player or computer) might feature less in improvisation or electro-acoustic practice than in the performance of a folk melody from memory. Moreover, some sound dimensions seem to be more susceptible to inclusion in an auditory mental image than others: in a sampling study of the occurrence of spontaneous musical imagery (Bailes 2006), I found that participants were more likely to image melody and song lyrics than harmony, timbre and expressive features of music. Intons-Peterson (1980, 1992) explored auditory imagery of environmental sounds in laboratory conditions and also found that timbral and dynamic dimensions seemed to be optional rather than essential to a conscious image. Harmony, timbre, dynamics and musical expression are, however, meaningful dimensions of musical experience. It would be wrong to strip down the notion of musical essence to those dimensions of auditory experience that can be mentally re-presented in consciousness. Nevertheless, translating from image to sound, it seems likely that musical dimensions important to the musical imagery of expert musicians would also be described as significant to the sounding of music. A content analysis of interviews with musical experts might reveal descriptions of musical dimensions important to image and sound.

Case studies

A semi-structured interview technique was employed that was designed to capture the individual experiences of the interviewees. While the respondents were selected because of their activities as professional musicians, the intention was not that they be considered representative of a wider musical population. The interviews reported here are necessarily limited by the musical and cultural backgrounds of interviewer and interviewees. Nevertheless, the focus on the experience of classically trained musicians represents a relevant viewpoint, coming from a tradition that promotes a key role for musical imagery.

The purpose of the interview was to explore emergent themes concerning the relationship of imagery to perception in musical activity and to allow the respondents to express their own accounts of their imagery experience. The interview schedule differed slightly for each respondent and was used as a prompt rather than a rigid set of questions. The aim was to encourage respondent-led discussion, albeit within loosely defined areas of researcher interest.

Interviews lasted between one and three hours. They were recorded with the permission of the respondents and later transcribed for thematic analysis. Each case study was analysed separately, examining the material repeatedly and systematically in order to determine emerging themes. Ultimately, a comparison of the common themes across the respondents was formally made (Bailes 2002). Common analyses relevant to translation of the image, across all three individuals, will be presented here.

Respondent profiles

Table 3.1 summarises the biographical details of each respondent. The broad areas addressed with each respondent are listed, along with the key emergent themes arising from an analysis of each interview.

Table 3.1 Respondent profiles

NG	
Biographical details	Composer.
	Teaches in the music department of a university in the United Kingdom.
	Experienced in directing contemporary music and in piano performance.
	Studied composition with Bernard Rands and David Blake.
Interview questions	Musical background, composition process, teaching, directing contemporary music.
Emergent themes*	Differences between imaging and imagining music, the respective limitations of imaging and perceiving music, conceptual flexibility, musical meaning, musical familiarity.

HP	
Biographical details	Pianist.
	Teaches in the music department of a university in the United Kingdom.
	Experienced in conducting.
	Studied with Cyril Smith and Nadia Boulanger, and worked with Messiaen.
Interview questions	Ideal working method of a performing pianist, including the importance of mental preparation, musical training, role as a teacher, aural skills.
Emergent themes*	Definition and centrality of a guiding 'conception', importance of formulating detailed thoughts about a piece before hearing it, balance in mental preparation, ideals and realities, conceptual flexibility, tricks to develop auditory perception.

WP	
Biographical details	Organist and pianist.
	Also experienced in choir conducting.
	Has congenital hearing loss and has been profoundly deaf since age seven.
	Works as a signer for the deaf in opera and musical performances. Runs a charity to assist in the musical education of deaf children.
Interview questions	How he perceives music, nature of his musical imagery, performing, conducting, signing.
Emergent themes*	Reliance on musical score, awareness and observation, perceiving through imaging, the 'working out' of music, memorising, imagination.

* Many themes emerged from the dialogue, but only those pertaining to the translation of musical imagery and perception are discussed here.

Interview analyses and discussion

Loss and gain

Translation from one language to another entails loss and gain. By virtue of no longer using the same word, expression or musical medium, some quality is lost, but there can be a gain through the new perspective afforded by the translation. In this way, Eco (2003) describes translation as a negotiation. For example, NG explained that translating sound to mental imagery inevitably involved a loss in the veridical nature of sound colour. For example, when asked to describe any surprising outcomes of finally hearing his compositions performed, he replied:

> I've always felt that the colour comes up more vividly in real life, and it's always better than you imagine it's going to be…Timbres are also going to be that much brighter—at least that's what I find—than the way I imagine them. They're always a little bit kind of, a little bit hazy, you know? They're not quite as resonant as they are in real life.

All three respondents expressed in various ways the importance of flexibility in perceiving and imaging music. In composition, this meant a delicate balance of fixing a musical idea without it becoming irremediably 'stuck' in notation or perception. In performance, the flexibility lay in the conception, a schematic framework that allowed for image and perception change. For WP, there was a need to ensure that learning from the technique of others did not transfer to mere copying without room for interpretative variation. Similarly, NG and HP described situations in which over-familiarisation with perceived music could occur: for NG this might involve hearing a composition repeatedly and becoming indifferent to its potential to grow, while for HP this would be the unwanted influence of other performers' renditions or being ruled by physical factors before a mental conception had been created.

NG believed that in composition, musical imagery could function to retain the balance of conceptual freshness. When talking about the 'use' of imagery in electro-acoustic composition (a medium in which he had only occasionally written), NG explained that having perceived the same music repeatedly could lead to over-familiarity with the material:

> An instrumental composer is imagining a performance or imagining a sort of idealised performance whereas the studio composer's hearing the real thing. And you can, in the studio, get bored with the material and feel that you need to produce more layers of activity to liven it up, whereas the listener coming to that afresh might find it very exciting.

It is true that a listener would rarely attend to a piece of music in as much detail as the composer. The composer's boredom from repeatedly hearing the material might be avoided by composing some of the work through imagery rather than perception.

Translation is more than mere transcription and requires imagination. There was general agreement as to the role of imagination in music making, with it differing slightly from imagery in its creative function of reaching beyond given material. Perceived or imaged representations of music might equally constitute this given material, so that imagination was related but separate. In composition, hands-on manipulation qf sound might be a more fruitful source of imaginative development, though imagery might also help break free from learned material through its inadequacies as an exact mental representation of given music. NG found the metaphor of a faulty tape to be a useful means of describing musical imagery. The weakness of mental imagery as a faithful reproduction lends it value for experimentation and modification. In this case, a certain gain in creativity is afforded by the use of imagery in composition rather than perception.

For HP, the 'imaginative impulse' in musical performance was at risk of being stifled by overly perceiving rather than imaging music. For him, going beyond the given meant looking beyond the surface information in the score, without distraction from the given material of perception, in order to create new meaning. In piano performance, HP described the principal advantage of forming mental imagery before physically tackling new music as freedom from technical constraints. It was suggested to him, however, that mental states were just as prone to becoming fixed as physical ones. When he first adopted the method of mental preparation and imaging sound before playing it, he said that he did become 'locked' into his imagery and he had since learned to rely more on the conception than the strict auditory image. He explained that conceptions could be refined and modified in accordance with changing ideas:

> The important thing that you don't want, or I don't want anyway, [is] a performance that is so completely planned and cut and dried in every detail that it has no possibility of further growth...I mean in a way a certain, if you like, imprecision has to be built into the business of being a musician.

HP goes to some lengths to balance flexible and stable dimensions of his performance approach by noting down his initial impressions of a piece for reference at a later date. The emphasis here, as elsewhere in his interview, was on using indirect methods to practise music, avoiding habit and maintaining a fresh perspective while deepening a musical understanding.

WP also spoke of imagination, as having had to establish an 'imaginative vocabulary' of timbre. Implicit in his use of the term is the notion of extending given material, as for him, timbral information is not presented aurally, but needs to be created through combining and extending proprioceptive, observed and analytical knowledge. Contrary to Marek's description of the oral tradition of learning the 'minutiae' of performance practice from listening to great performers, the two expert performers interviewed in the current research described quite different approaches to image formation. For HP, hearing examples of music he wished to perform was to be avoided in preference for a highly autonomous mental image of the music. For WP, musical perception concerned visual perception and hands-on experience of playing an instrument.

Mediating the musical 'essence'

Translating image to sound and sound to image implies at least a basic level of equivalence between the two phenomenal experiences. For example, where there is an intention to communicate musical imagery in sound, it can be hypothesised that the composer/performer will translate the most meaningful elements of the mental image. The elements reported to feature in NG's mental imagery seem to correspond with those meaningful to his concept of music. In fact, he deliberately pointed out this relationship, saying that musical imagery necessarily related to an individual composer's musical language. An example of this concerns what he considers to be good examples of timbral writing in the composition of others: '[i]t's partly to do with working with the instrument and getting the best out of the instrument…finding something which is so characteristic of the instrument that you can't imagine another instrument playing it…Some timbres seem absolutely right for particular situations.' NG reported a conversation he had with a colleague about whether harmony was experienced as part of the mental image for music, using the example of the slow movement from Rachmaninov's *Symphony No. 2 in E Minor*. His argument was that harmony was necessarily 'heard' as integral to musical meaning and his colleague's disagreement made him 'question his musicality'. When asked if there were any musical styles that he would find harder to internalise and remember, he explained that he could not create an image of the whole of *Gruppen*,[1] but that as gesture was the main meaning of that music, so gesture constituted the aspects of sound he could hear in his mind's ear.

Essential to HP's musical understanding and method of approaching performance is the 'conception'. This can take the form of a mental musical image, but bears closer resemblance to a more abstract schema generated for each piece. Indeed, the conception might be understood to be a generative force behind an individual piece of music: essential to its musical understanding and able to accommodate surface-level changes. He compared the conception with the largely structural overview held by a novelist. Once clear in the mind, the conception acts as a form of imagined model against which to compare all perceived renditions. HP explained that 'I like to know what I'm aiming at, and then I will find the means to realise it'.

For HP, it is important to maintain the ideal conception. For example, he described his reluctance to practise at the venue for his next concert because of the piano's inadequacies, which might tarnish his idealised image of the music. The physical inadequacy of the instrument is largely beyond his control, while the ideal conception of how the performance should go is not. While an ideal performance is difficult to achieve, 'you're never going to get it unless you've imagined it. But you'd be amazed what you can do if you have actually set Everest in front of you…[because] the other thing is you're motivated by your imaging. You want it. You're not satisfied with less.' This was in contrast with the neglect that HP observed in most people's inner hearing. He bemoaned what he described as the 'sticking plaster' approach to solving performance errors, whereby individual notes were physically repeated, rather than listening and imaging the desired sound before attempting a performance: 'I suppose the huge lesson that I learned from my teacher…Cyril Smith, whose thinking in this area was incredibly advanced, which is that…we want to lead our hands.'

As a professional pianist, HP was concerned with the memorisation (formation of a memory image) of music. He spoke of his approach to memorising music as a developmental trajectory. For him, rote learning was something suitable to a child's mind, but the more experienced a musician became, the more an understanding of musical meaning was important to the process of remembering. As the number of musical episodes that have been experienced increases, the more the semantic rather than the episodic quality is significant. Accordingly, HP increasingly found the semantic musical meaning and conception to be important in learning a piece of music. The implications of this are an increasingly analytical relation of the music to abstract musical knowledge.

WP described harmony as a salient musical feature, as it not only featured in his favourite musical moments, it seemed to elicit the strongest emotional response. When describing favourite moments in pieces of music, WP's description took the form of the analysis of modulations. Speaking of *West Side Story*, he says, 'I never get tired of listening to that. It's such an incredibly analytical piece of music.' Analytical structure features heavily as the essential component in describing his method of memorising a work. His musical memory is more structural than note level: 'When I say that I memorise a piece of music, I obviously don't memorise every single note…I suppose it's a pared down representation of the whole thing…I suppose I'm really aware of form in a way.'

It is impossible to know to what extent WP's aural imagery resembles that of a musician without a hearing impediment. In spite of his emphasis on an analytical grasp of score, he also describes his music in terms of an 'aural tape' playing in his mind. When asked whether he envisaged the score mentally when memorising music, he replied, 'No, I don't actually…It's like I have a tape in my head, and I'm playing the tape.'

Schemata based on past experience mediate perception, so that each perceptual moment is a process of extracting information from the environment that is meaningful to the individual. The musicians in these studies described imagery comprising features that correlated directly with those features they found interesting and meaningful in perception. This is not unusual when considered alongside the reported imagery of some famous composers. Stravinsky continually stressed the important use of instrumental colour as an integral part of a musical idea. He presents the way to write effective instrumentation as comprising imagination and declarative knowledge about the instrument, stating that Berlioz had both aptitudes (Stravinsky and Craft 1979:29). Elgar also experienced instrumentation as part of the musical meaning of his composition:

> The fact is I mentally hear the instruments, and when scoring put down what I feel that the sentiment of the words, if there be words, demands for the most perfect expression attainable. So far as I am concerned the thing is already complete in my mind: to make others feel it as I do is the trouble. If I could only write as fast as I think! (Quoted in Buckley 1912:87–8)

The translation process

The notion of balancing loss and gain in translation has already been outlined. All three interviewees described intentionally exploiting this property of the translation process to suit their needs. When questioned about the relative merits of writing down compositional ideas or guarding them as mental constructs, NG clearly articulated a need to balance fixing ideas and maintaining flexibility. Describing musical imagery, he explained:

> If it's a playback mechanism you can't actually be certain that every time it's the same, and I think this is the beauty and the drawback to notation that it actually fixes a version. If you're unlucky, that version becomes something that you can't develop any further because it becomes so fixed you can't see any potential in it any more. It's definitive; it's complete. Whereas, you know, sometimes it's more valuable to try to keep the ideas fluid.

NG describes imagery in his composition less as a static mental copy of sound than a way to free the imagination in its departure from repetition and fixed musical detail. A 'casual listener' to his own musical imagery, NG might attempt to 'capture the drama':

> I've got strategies for trying to get the music down quickly. I mean there are two distinct processes: one which is to try to write down the music that's going on in your head, and the other which is to project more music. So one is constructivist, if you like, and the other is trying to capture something on the wing.

Giving an example of 'capturing music on the wing', NG described how in order to produce a 20-minute composition assignment as a student, he had once attempted to imagine music in real time for that period, but

> [w]hat came out eventually at the end of that summer was richer than what I heard in my head. Because I hope that I built into it layers of meaning…layers of activity which the casual listener the first time round wouldn't have heard. And in that position I was of course the casual listener.

The composition process necessitates making something more of this inner sound, so that the product is greater and richer than the image. Composers might only rarely experience moments of inspiration in the form of a complete auditory image. Rather, an amalgamation of auditory knowledge, based on perceptual experience and the unconscious association of ideas, is likely to underpin the bulk of musical creation. Where a conscious image of sound can be of particular value is to present auditory information to the mind's ear for contemplation, selection or transcription.

HP envisaged the translation from conception to surface detail to be a continuing cyclical process, described as follows:

> I think you start with a conception, but clearly when you work on the details, I mean they are conceived too, and will feed back into the wider conception which will be modified accordingly, and indeed they may be modified by the instrument which you're playing on, and even in performance itself by the performance, and so on.

Familiarity with music, through its repeated perception, is assumed to lead to a strong corresponding mental image. An ideal for HP, however, is to avoid excessive familiarity with the music he is working on. The reality of over-familiarity is described as a loss of freshness. A direct comparison can be made between this performer's desire to rejuvenate his musical ideas and NG's balancing of perception and imagery in order to maintain creative newness. For HP, over-familiarity in motor terms means that 'if you're so conditioned in your nervous system and so on, doing it just like this, you can't react anyway'.

HP's extremity of isolating the mental preparation of a performance should be contrasted with his description of the balance needed to become what he described as 'a great pianist':

> Everything has to be in balance. When things are in balance you're playing well, and when you're not playing well, when you've lost your form, whatever it is, it's just like golf or something like that, or tennis…it's usually almost always because there's some key area that you're neglecting, whether ear, or indeed the conception.

This balance relates to the practicalities of managing repertoire as mental and physical preparation:

> It seems to me in a way that every day you come to practise a piece, and if it's a very difficult piece you may have to practise it day after day after day, but still every day you have to bring some new imagination to it, because otherwise the work simply becomes kind of technical, and in some way divorced from the imaginative impulse.

HP also mentions balance as important to the work of a conductor in safeguarding a conception and accommodating any change brought about by individual performers.

> I say the task of being a conductor is really…treading that particular tightrope, isn't it?…There has to be some kind of balance in a conductor's work, and so it is obviously with any kind of instrumentalist, there's a balance between your pre-rehearsal planning, or pre-practice planning, and then what you actually do. But essentially, of course, you're able to diagnose, correct, improve and so on because you do know what it is that you want.

The balance between mental and physical dimensions might change during the process of preparing to perform a piece, but the guiding conception, a fundamentally mental measure, is present throughout. HP described his own approach to performance as 'controlled imprecision', a phrase that exemplified the balance required between conscious and unconscious thought in a piece. Control needs to be introduced and this is a predominantly conscious phenomenon. Unwanted unconscious influence should be avoided: 'The fact that we may be unconsciously influenced in the early stages by what feels comfortable, or safe, or whatever, is bad.' For HP, it is a lesser concern that processes become automatic than that processes should result from mental thought rather than physical tendencies.

Holding an idealised performance in mind requires imagery, but acting on it requires a perceptual feedback mechanism. HP said that ideally, the formulation of a conception, and the subsequent ability to image the desired sound, would allow a performer to diagnose their own problems. 'It is unquestionably the only path towards radical improvement, and huge standards…I am convinced it is the only method…if there's a secret…Because it enables you to be your own teacher, your own doctor. You diagnose everything yourself.' Based on his own experience, HP outlined useful tricks for a performing pianist to develop their perception and imagery—for instance, anything that involved playing the music in a physically different way, such as hand swapping, would necessitate aural rather than kinaesthetic cohesion. Singing the more aurally insecure left hand of a piano part while playing the more melodic right hand would also guarantee a thorough aural knowledge of the music.

The translation process of sound to image and image to sound described by WP is unique to his experience as a deaf musician. He was asked whether he could imagine the different sound qualities of different instruments when he read the score. His response was positive and he was able to explain it in terms of firsthand experience in playing different instruments when he was younger. He spoke of having had the 'chance of feeling' the bass clarinet. Interestingly, he expressed a particular interest and ability in orchestrating music: 'Now by rights I shouldn't be able to do that, 'cause you need a good ear…You need to be able to tell what instruments will sound like when you combine them…I freely admit I didn't have an extensive knowledge of the workings of these instruments.'

WP attributes his ability in orchestration to a vivid imagination, explaining, 'I've tried playing different instruments, or being in close proximity to them you build up this imaginative vocabulary, for want of a better word. And somehow your brain just absorbs it.' Thus, imagination, more creative than imaging, plays a crucial role in the translation of timbral experience for WP.

When asked whether there had been any overt emphasis on imaging music in his musical training, WP replied that hadn't been the case, probably because he had relied on performed examples to learn. Really performing music forms a central part of WP's experiences. He has learned to play the organ through observation, yet he says:

> Interestingly enough, I don't frequently have the image of somebody physically playing. If I look at a piece of music I don't imagine somebody sat [sic] at a keyboard and how they're going to finger it…That could be because every performance that you do has got to be unique in your own interpretation.

This final comment is illuminating, as it demonstrates flexibility required to accommodate individual interpretation. When asked whether WP deliberately listened to and attended the performances of others when preparing a piece for performance, he explained that he tended to imagine his own interpretation.

WP frequently described a process of 'working out' some aspect of music. He said, for instance, 'I don't find it difficult to work out the harmony and even the structure just from looking at it' and 'I prefer to sit down and work out my own'. WP relies on musical score in order to experience music, though he is unable to explain how he acquired his apparently detailed auralisation skills. Score reading combines with a general awareness and observation of musical situations, such that music for this organist is more clearly an amalgamation of sensory modalities than otherwise acknowledged in music perception. He also describes perception in imagery terms—for, to him, perceiving necessitates image formation and the two are part of the same process.

Fluency within a given 'language'

Successful translation requires fluency in both languages. The accurate translation of sound to image or image to sound equally entails fluency in both. Interviewees expressed different comfort levels with imagery and perception relative to their experience or practice in each. When asked whether practice in composition had affected his own image formation, NG replied, 'Yes, I think it must…I can't say that I'm tremendously confident that I'm better at it now than I was, but I suppose I must be.'

Practice in musical imagery for HP meant balancing fixed and fluid tempi, as he felt that performers should be able to image tempo 'and in a way be able to switch ideally effortlessly from one to another, from say a very fast tempo to a very slow tempo, and vice versa'. Freedom is an important theme for HP. In general, he reports that his mental imagery is not restricting but liberating, and that 'the marvellous thing about imaging is that I can fly anywhere I want to'. From that he goes on to say that physical limitations, time and ambition are the only factors that could prevent the communication of a perfect musical interpretation: 'I've no doubt that you could put…a lot of repertoire in front of me, and that I can inside myself imagine what Smith would have called a world-beating performance.' HP made it clear that for him, score reading was a more productive exercise than sight-reading at the piano, as this bypassed technical problems in favour of a well-founded musical interpretation. For this to be the case necessitates exceptionally fluent imaging skills.

The notional division between sound and image is irrelevant to WP. He explained that he relied entirely on the printed score, as listening to music 'doesn't mean anything, its just some abstract noise with no recognisable shape, no recognisable form'. He went on to exclaim, 'I just cannot imagine life without a score'. For WP, score is the primary musical experience. How is WP able to translate score into a musical image given his lack of conscious aural memory and having been profoundly deaf from early childhood?

> I never knowingly sat down and taught myself how to read a score. I always thought that that was something that anybody could do…It's not something I've knowingly learnt to do. I don't know whether that's been some sort of automatic compensation for my hearing loss, or the fact that I have to relate to what I see.

WP explained that scores differed in their difficulty of reading and subsequently in imaging them. Of a particularly dense score, he said, 'that's not to say I couldn't do it. If you gave me time I could sit there and work out what was going on.' His approach when presented with new score was to 'flick through it' to gain an overview and pick out important features. Given that the musical information represented in score is WP's primary access to music, it is unsurprising that an incomplete score is troubling to him:

> If somebody gives me a lead sheet and some lyrics I don't feel confident at all. I find it so hard to do. *Blood Brothers* is an example. I've been signing *Blood Brothers*...for about six years now. They don't have a score. All I have is a vocal selection book so a vast amount of the time I've no idea what the band is doing.

WP reported difficulties in recreating music he had heard without a score. For instance, as a child, he heard Mahler's *Symphony No. 8 in E Flat Major* and only subsequently bought a recording and a score. He was then unable to re-experience the passage he had liked at first hearing, perhaps due to memory problems or perhaps because of a faulty original perception: 'What I imagined was happening and what was really happening may not have been the same thing.' WP proposed that a mismatch between perception and image could have occurred. He expresses this idea explicitly:

> If I go to a concert or on the rare occasions I might listen to a piece of music without a score, but I do know what's going on, I can actually conjure up some kind of picture in my head, but when I get the score afterwards I find it bears no relation at all to what I thought I heard.

This is an example of perceiving music through imaging. Comparing a mental image with the score is, for WP, tantamount to comparing his mental image with a 'listening'.

Conclusion

The purpose of the case studies reported in this work was to gather detailed information about the perspective of individuals on the relationship of imagery to perception in professional musical life. Their experiences have been presented as forms of translation from image to sound and sound to image. The reader should note that in addition to the translational level of musical thought to sound, or sound to mental re-presentation, a number of interpretative levels have been superimposed on the interviewees' experience. Not only has the author translated the experience of the three musicians according to themes of translation, the interviewees were asked to reflect on how introspection and self-reporting affected the very experience of musical imagery during the interview process. NG commented on this and, as he did so, he described imaging *L'Après-Midi d'un Faune*, saying:

> I suspect that it's to do with the sense of what in the textures you're listening to are most immediate...And I would suspect that trying to describe it to you would tend to make these layers drop away. So what's

immediately retrievable will stay—so the flute will stay and the harp will stay at the beginning—but other things might drop away.

This presents imagery in the light of an ability to attend to different sounds much as might occur in perception. It is important as an indicator of how changes in introspective attention might alter the entire imagery process and represent an interpretative translation in themselves. This is a facet of a further level of translation described in the introduction—namely, the translation in musical process between unconscious and conscious thought.

The chapter began by asking how conscious musical thought, or musical imagery, translated to musical production. It is not within the scope of this work to consider the physical coordination involved in sound production. Instead, this research hopes to highlight the role of auditory imagery in planning musical production, and conversely as a translation from the perceptual experience of sound. Imagery and perception could be so intertwined as to appear inseparable. Certainly, for WP as a deaf musician, perceiving music necessitates the immediate formation of a musical mental image and imaging music through score reading constitutes perception. The musicians of this study, however, also drew certain fairly clear boundaries between image and sound. In translation terms, it is beneficial to retain a certain fluency in both 'languages', as each is essentially linked to an underlying conception or schematic framework, but each provides a complementary interpretation, unattainable in isolation. Imagery as described by the respondents might be an idealised musical representation, being held up as the perfect goal in performance. Perception might never live up to the exemplary image. This perspective imbues imagery with a freedom that cannot be matched in musical production. At the same time, imagery is acknowledged to be a flawed representation of perceptual experience.

In translating from one language to another, there is inevitably a negotiation between loss and gain. Saxton (1998:6) articulates the intricate negotiation of perception, imagery and imagination involved in composition, saying that '[f]or a composer, there exists continual two-way osmosis between the material itself and applied methods of treating the "received" musical ideas'. We have seen that perception or imagery can furnish these 'received' ideas, while imagination works to develop this, either consciously or unconsciously, to present in a new form. Perception and imagery are described as playing crucial roles with respect to the need for creative musicians to fix their ideas. This applies as much to performers as to composers. For instance, in accordance with the views of HP, Brendel (1976) says that when music is not easy to retain in memory, having to relearn it contributes to a fresh performance. Here the mental image is not a fixed 'master record' but is closer to a flexible conception. Musical imagery and musical sound might be expressed as translations that serve to 'augment the significance and expressivity' (Eco 2003:82) of musical experience. It is certainly difficult to conceive of the existence of one without the other.[2]

References

Bailes, Freya 2002, Musical imagery: hearing and imagining music, PhD dissertation, University of Sheffield, Sheffield.

Bailes, Freya 2006, 'The use of experience-sampling methods to monitor musical imagery in everyday life', *Musicae Scientiae*, vol. 10, no. 2, pp. 173–90.

Brendel, A. 1976, *Musical Thoughts and Afterthoughts*, Robson Books, London.

Brodsky, W., Henik, A., Rubinstein, B. and Zorman, M. 1999, 'Inner hearing among symphony orchestra musicians: intersectional differences of string-players versus wind-players', in S. W. Yi (ed.), *Music, Mind, and Science*, Seoul National University Press, Seoul, pp. 370–92.

Buckley, R. J. 1912, *Sir Edward Elgar*, Second edition, John Lane, London.

Collins New English Dictionary 1997, HarperCollins Publishers, United Kingdom.

Eco, Umberto 2003, *Mouse or Rat? Translation as Negotiation*, Weidenfeld & Nicolson, London.

Godøy, R. I. and Jørgensen, H. (eds) 2001, *Musical Imagery*, Swets & Zeitlinger, Lisse.

Harvey, J. 1999, *Music and Inspiration*, Faber and Faber, London and New York.

Intons-Peterson, M. J. 1980, 'The role of loudness in auditory imagery', *Memory and Cognition*, vol. 8, no. 5, pp. 385–93.

Intons-Peterson, M. J. 1992, 'Components of auditory imagery', in D. Reisberg (ed.), *Auditory Imagery*, Lawrence Erlbaum Associates, Hillsdale, NJ, pp. 45–71.

Kvifte, T. 2001, 'Images of form: an example from Norwegian *Hardingfiddle* music', in R. I. Godøy and H. Jørgensen (eds), *Musical Imagery*, Swets & Zeitlinger, Lisse, pp. 219–35.

Marek, G. T. 1982, 'Toscanini's memory', in U. Neisser (ed.), *Memory Observed: Remembering in natural contexts*, W. H. Freeman and Co., San Francisco, pp. 414–17.

Mountain, R. 2001, 'Composers and imagery: myths and realities', in R. I. Godøy and H. Jørgensen (eds), *Musical Imagery*, Swets & Zeitlinger, Lisse, pp. 271–88.

Penrose, R. 1999, *The Emperor's New Mind: Concerning computers, minds and the laws of physics*, Oxford University Press, Oxford.

Retra, J. 1999, An investigation into the musical imagery of contemporary composers, MA dissertation, University of Sheffield.

Robertson, P. 1996, *Music and the Mind*, Channel Four Television, London.

Saxton, R. 1998, 'The process of composition from detection to confection', in W. Thomas (ed.), *Composition, Performance, Reception: Studies in the creative process in music*, Aldershot, Ashgate, pp. 1–16.

Seashore, C. E. 1938, *Psychology of Music*, McGraw-Hill, London.

Stravinsky, I. and Craft, R. 1979, *Conversations with Igor Stravinsky*, Faber Music Ltd, London.

Endnotes

[1] *Gruppen* (1955–57) is a piece for three orchestras by Stockhausen. The instrumentalists surround the audience on three sides and the music is characterised by the superimposition of independent tempi.

[2] The research reported in this chapter is adapted from a portion of my doctoral work conducted at the University of Sheffield, United Kingdom, and supported by an Arts and Humanities Research Board scholarship. I wish to thank my interviewees for their generous participation in this research, as well as Karen Burland, Eric Clarke and Hazel Smith for comments on previous versions of this manuscript.

4. Translating the tradition: the many lives of *Green Bushes*

Jennifer Gall

Audio samples relating to this chapter are available online at:
http://epress.anu.edu.au/titles/sounds_translation_citation.html

Transcribed below are the words of *Green Bushes*, as sung by Sally Sloane in Australia, to which I refer throughout this essay.[1]

The Green Bushes

(Sung by Sally Sloane and recorded by John Meredith in 1956)

As I went a-walking one morning in spring,
To hear the birds whistle and the nightingale sing,
I spied a fair damsel, so sweetly sang she,s
'Down by the green bushes where 'e thinks to meet me.'

'Oh what are you loitering for, my pretty maid?'
'I'm a-loitering for my true love, kind sir,' she said.
'Shall I be your true love, and will you agree
And forsake the own true love and go along with me?'

'Oh come let us be going, kind sir, if you please,
Come let us be going from under those trees.
For yonder is coming my true love, I see,
Down by the green bushes where he thinks to meet me.'

But when he came there and found she was gone,
He looked all around him and cried quite forlorn,
Saying, 'She's gone with another and quite forsaken me,
So adieu to those green bushes for ever,' cried he.

'Oh I'll buy you fine beavers and fine silken gowns,
I'll buy you fine petticoats flounced to the ground,
If you'll prove loyal and constant and free,
And forsake your own true love and go along with me.'

'I want none of your petticoats nor your fine silken walls/robes [printed as robes, sounds like walls][2]
I was never so foolish as to marry for clothes,
But if you'll prove loyal and constant and true
I'll forsake my own true love and go along with you.'

[Last verse from the other three field recordings]
'I'll be like some schoolboy, I'll spend all my time in play,
For I never was so foolish as to be lured away.
No false-hearted young girl shall serve me so any more,
So adieu to those green bushes, its time to give o'er.'

The transmission of folk songs provides a continuous demonstration of sounds in translation. In the early twentieth century, the process of oral transmission of traditional music was revolutionised with the use of recording devices such as the phonograph adopted by Australian-born Percy Grainger. This recording activity marked the translation of the music of a small community into the musical language of a much broader audience[3] not obliged to listen to the song by geographical and social origins or emotional connections with the singer. By mechanically recording folk songs and translating the melodies into notated art music, Grainger was separating the living folk song from the ritual transmission of the oral tradition. The melody remains recognisable as the folk song, but the distinctive stamp of the original singer disappears. While Grainger's method demonstrates one process of sounds in translation, I am interested in the relevance of folk song as a reflection of the minds of its singers, an investigative process that also acknowledges the use of sound recording in contemporary folk song transmission. This chapter examines why and how I have used recording technology to investigate the rituals[4] of performance connected to domestic routines associated with the oral tradition, and to participate in the continuing life of the song through my own recorded performance. There are three parts: the first is an exploration of the historical sources of *Green Bushes* and the symbolism in the song; part two examines Sally Sloane's versions of the song in relation to James Porter's (1976:7–26) 'conceptual performance model', in which he states, 'Whatever "the song" is, its identity cannot be demonstrated, nor other features such as its existence through time fully delineated, until we are able to trace that identity in the mind of the singer or a number of singers' (p. 11).[5] Part three relates this theory to my recording of *Green Bushes*, in which I seek to integrate the influence of the singer I learned the song from (Sally Sloane) into my version.

Part one

From the earliest printed sources, about 1816, *Green Bushes* crossed and recrossed boundaries of musical genre and culture, from the oral tradition to the popular ballad press, from the theatre into Western art music and carried across the world through the oral tradition to continue its life through the lives of its singers in Australia. *Green Bushes* was documented (<http://www.bodley.ox.ac.uk/ballads/ballads.htm>; <http://www.csufresno.edu/folklore/Olson/index.html>; <http://www.folkinfo.org/topic.asp?topic_id=301>) in field recordings and notations of the oral tradition in Ireland, England, Canada, Nova Scotia and the United States, and it was published in many broadside ballad versions in the eighteenth and nineteenth centuries in England and in New York. Mrs Fitzwilliam sang it in John Baldwin Buckstone's play *Green Bushes* (1845) and the performances of this play in England, Australia and America either revived the existing song or encouraged transmission of the new version as it was sung in the play. The Bodleian Library Broadside Ballad *Sweet William*, with instructions directing that it be sung to the tune of *Green Bushes*, was printed between 1813 and 1838, and *Among the Green Bushes, Catnach (London), 1813–1838* (<http://www.bodley.ox.ac.uk/ballads/ballads.htm>) predates the use of the song in Buckstone's play. Peter Kennedy in *Folk Songs of Britain and Ireland* (1975:378) gives six field-recorded versions and 26 printed versions, some of which are transcriptions of field recordings or notations. Taking into account the evidence of the broadsides and other related versions of the song—*The Cutty Wren*, *The Queen of May*, *The Shepherd's Lament* and *Sweet William* through either melody, theme or related lyrics—and the many Broadside cross-references to songs related to *Green Bushes*, I would agree with Kennedy (1975:378) that 'we can presume it was traditional long before it was used by Buckstone'.

Documentation for published versions of the song is comprehensive, implying that it enjoyed wide popular circulation and performance. *Green Bushes* has engaging narrative complexity and there is great potential for the singer to invest their personality in the performance. The story engages the listener and, at a deeper level, the narrative links the audience to the song through the characters, who represent archetypes,

> 'felt' as embodied forces, pulling this way or that, intersecting points in a network of linked relationships, driven by passion to their destiny. By comparison with the invisible but strongly felt network that links them and the Fate that controls them their personalities do not matter. (Muir 1965:157)

The protagonists' lack of personality allows the listener to empathise with the emotional web acted out in the course of the song by the male and female characters.[6] Symbolism in the song connects the singer and listener to the past. *Green Bushes* epitomises the ability of orally transmitted folk music to enable singer and listener to experience a sense of belonging to an all-encompassing pattern of fate and relationships, while simultaneously recognising that the song

63

is a reflection of these patterns. Singer and audience ideally form a community of shared meaning in the course of the performance.

It is a circular narrative. A man out walking in spring sights 'a fair damsel' singing 'Down by the green bushes where he thinks to meet me'.[7] In Sloane's version, the woman sings a song that attracts the new suitor ahead of the old one: perhaps a test to see who arrives first? Possibly these are references to some other, earlier layer of magical meaning and ghostly, revenant lovers, which have been simplified over time. He asks what she's waiting for and she tells him it is her true love; he offers to take the place of the man she's waiting for. When she sees her old lover coming, she leaves with the protagonist. The lover arrives and realises he's too late. Meanwhile, the first man offers the woman fine clothes to entice her to commit to him, but she explains that all she wants is his fidelity. In the final verse, the old lover announces himself cured of the attachment to the woman. In the melodically related broadside, *Sweet William*, the woman is rewarded for her faithful wait by her returning sailor lover with many rich and exotic gifts. Is the girl really 'false hearted' or is she simply tired of waiting for a lover who would prefer to be off adventuring? Or has her 'true love' died and is returning as a revenant lover?[8] This is a possible interpretation given the emblematic meaning of the nightingale[9] mentioned in verse one as a messenger from the dead or transformed loved one, and the association of the colour green with death and the supernatural. The symbolism in the song offers the possibility of many different interpretations.

The colour green has numerous symbolic meanings. 'Green language' and 'language of the birds' are associated historically with 'oblique writing styles used by alchemists and other mystical initiates' (<http://altreligion. about.com/library/glossary/bldefgreenlanguage.htm> Stewart 1977) and this symbolism could have played a part in the early evolution of the song. Certainly, the ambiguities in the narrative suggest the possibility of changeable meanings. Green is a colour associated with the dead in ballads (Wimberley 1965:242) (for example, *The Wife of Usher's Well* and *Green Gravel*) as well as with virginity, deception and disguise adopted to test a lover's fidelity. At the most fundamental level, even if unaware of the traditional meaning of these symbols, singer and audience are able to connect these references to their own experiences through imagery that evokes memories. It is the ambiguity of folk song narrative that ensures its appeal to a wide range of listeners and singers. The symbolic references and the melodic patterns of *Green Bushes* convey the idea of a recurring pattern of relationships.

Grainger collected two versions of *Green Bushes* in England, as part of an extensive, pioneering field recording to notate disappearing traditional English folk songs. He was inspired to use technology to capture the subtleties of ornamentation and phrasing peculiar to the oral tradition of singing English folk music by Madame Leneva's notations of Russian folk-art songs collected with a phonograph at the turn of the century (Slattery 1974:43). Grainger wanted to exploit the extraordinary wealth of musical ideas to be harvested from rural England for translation into Western art music. His translations of these folk songs helped popularise the music in a way that altered the originality and spontaneity of the version recorded from his informants. Writing in 1908, Grainger described his English folk song collecting experiences:

> When I first started collecting folk-songs with the phonograph, in the summer of 1906, in North Lincolnshire, I was surprised to find how very readily the old singers took to singing into the machine. Many of them were familiar with gramophones and phonographs in public houses and elsewhere, and all were agog to have their own singing recorded, while their delight at hearing their own voices, and their distress at detecting their errors reproduced in the machine was quite touching. (Grainger 1908:147)

Grainger was describing the transition of music that evolved as part of a discrete community's musical life performed by a soloist for her/his community into traditional music performed and rearranged in a manner that was never envisaged by the original 'composers' or traditional performers. Suddenly, because the performance was captured by a recording apparatus and became replayable, 'errors' mattered. In a live performance tradition, deviations from the 'perfect' rendition of the song were and are the life of the song. Grainger also viewed these songs as common property, available for translation by anyone into another musical form. On at least one occasion, he hid under an old woman's bed to record a song he particularly wanted, knowing full well that she regarded it as her exclusive possession (Slattery 1974:48). Unlike Joseph Leaning and Sally Sloane, who learned their version by listening to another singer, then made the song their own, Grainger tried to make the song the property of all listeners through translation into the much broader, de-personalised language of Western art music.

This is a sample of Grainger's recording of Joseph Leaning singing *Green Bushes* (refer to Audio 4.1) and its lyrics:

> As I was a-walking one morning in May,
> To hear the birds whistle and see the lambs play.
> I beheld a fair damsel so sweetly sung she,
> Down by the green bushes where she chanced to meet me.
> 'Come let us be going kind sir if you please,
> Come let us be going from under the trees
> For yonder he's coming, he's coming I see,

Down by the green bushes where he thinks to meet me.'

(Leader LEA 4050 LP, mono UK, 1972)

Grainger's work, *Green Bushes (Passacaglia on an English Folksong)*, repeats the verse for eight minutes while the orchestral accompaniment becomes increasingly extravagant. The elaborate instrumental arrangement represents a paradox given Grainger's love of the unaccompanied vocal tradition he was recording and his declaration that '[t]he old folk-singers were not limited to the harmonic poverty of instruments' (Slattery 1974:45). For all Grainger's attention to the detailed ornamentation and phrasing that made *Green Bushes* significant within the context of the oral tradition, these subtleties have disappeared in his orchestral arrangement. Perversely, Grainger based his setting 'on a version of *Green Bushes* noted by [Cecil Sharp] from the singing of Mrs Louie Hooper of Hambridge, Somerset' and used his own recording 'to a lesser extent' (Lewis 1991). Grainger believed that the rhythmic structure and repetitive melodic pattern of the song connected the music to the past. This was the essence he sought to capture and translate into his own work, but it was not the kind of transmission operating as part of the oral tradition. In his program note for a performance in 1930, he wrote:

> *Green Bushes* strikes me as being a typical dance folk-song—a type of song come down to us from the time when sung melodies, rather than instrumental music, held country-side dancers together…In setting such dance-folksongs…I feel that the unbroken and somewhat monotonous keeping-on-ness of the original should be preserved above all else. (Lewis 1991)[10]

Perhaps the greatest appeal of the song was its ambiguous modality, enabling Grainger to overlay 'a harmonic treatment covering a range of seven or more keys' (Lewis 1991).

Music of European origin, such as *Green Bushes*, remains a strong vein in Australian folk music, and the variations that occur here reveal much about the evolution of a distinctive folk music. The transmission of folk song in Australia represents exchange between men and women and men and men, through relationships such as father and daughter, wife and husband, mother and son as well as sharing songs between networks of friends, receiving music by post in remote locations and learning new items from travellers, in the way that Sloane learned *Green Bushes* from Jack Archer, an itinerant railway worker.

Part two

Green Bushes is an important example of the introspective ballads that contrast with the more popular, diversionary Australian folk songs such as *The Rybuck Shearer*, *Click Go the Shears* and *Botany Bay* favoured in the bush band repertoire.

Like Catherine Peatey's versions of *The Bonny Bunch of Roses* and *The Female Rambling Sailor*, they provide not only a continuing link to a European cultural background, but a framework adaptable by the singer to express their own emotions and to reflect their personal place in patterns of relationships. Folk song, like myth,

> no matter how ancient its origins or its subject matter, is always concerned with contemporary relationships, here and now…its value lies not in its truth to any actual past whose reality we can establish or disprove but in its present usefulness as [a] guide to values and to conduct. (Small 1998:100)

The existence of *Green Bushes* in the repertoire of Australian singer Sloane demonstrates how life circumstances and relationships between the singer and another musician result in transmission and survival of the song. These living links demonstrate the operation of tradition as defined by the Australian scholar Barry Macdonald:

> a) shared repeatable activity or complex of activities…and b) the activation of a certain spiritual/emotional power in the relationship-network of those involved in the collaboration. This power is produced by, and in its turn, generates the conscious desire for the activity, its objects (for example, particular songs, styles or stories), and the relationship network itself to persist—just as they had in the past, so on into the future. (Macdonald 1996:16)

Tradition enables a singer to learn a song from the past and send it on into the future through a performance provoked by feeling, and the transmission of the song *Green Bushes* demonstrates this process. James Porter's (1976; see also Gower 1968) 'conceptual performance model', which he designed to analyse Jeannie Robertson's singing of the Scottish ballad *My Son David*, is a particularly useful tool for understanding this pattern of folk song transmission as sounds in translation. Porter proposes that traditional singers choose certain songs as favourites because they deal with issues that are important in their lives, consciously or unconsciously. Their performance of the song reflects this in subtle ways.

> [T]he singing of a ballad cannot be viewed simply as an act in which a text (understood to mean a 'story') is set in motion by a singer with a tune, but as a complex, existential process in which units of both cognitive and affective experience are embedded. (Porter 1976:17)

The following discussion of Sloane's version of *Green Bushes* is based on four different performances recorded over a number of years by two collectors:

- 1953 NLA TRC 2539/5 recorded by John Meredith
- 1956 NLA TRC 4/19 recorded by John Meredith
- c1957 NLA TRC 2539/79 recorded by Edgar Waters[11]
- c1960 NLA TRC 4/13-14 recorded by John Meredith.

Sloane's version of *Green Bushes* is consistently slower and more legato than the Leaning version recorded by Grainger: '[h]allmarks of this performance are the pure, unemotional vocal style, long lyrical phrasing and the relaxed tempo that draws the listener in to the story' (Waters 1957:6, 1957:3). In fact, the tempo, while rhythmically flexible, is consistent with that of a tactus or resting heartbeat of roughly 63 beats a minute, a fundamental reminder of the interrelationship between singer and song. In the field recordings of *Green Bushes*, melodic subtleties are not obscured by instrumental accompaniments and imposed arrangements. Modality and vocal nuances are evident in the recordings. As manuscript notation cannot capture the subtleties of rhythmic variation, or the textural effect of Sloane's phrasing, the best illustration of these characteristics is a sound sample of the longest six-verse version of *Green Bushes*: TRC 4/19 recorded in 1956 by John Meredith (see lyrics above).

I have endeavoured to gain an understanding of the connection between Sloane's life and the words of the song—what Porter (1975:9) describes as 'understanding the musical process in formation, from the inside as well as the outside'. Meredith and Anderson (1979:173) noted that Sloane learned *Green Bushes* from Jack Archer when she was twelve years old. The song remained a favourite with Sloane in her adult life and was recorded by several different collectors who visited her. *Green Bushes* was important to her for many reasons. First, Archer must have been a significant personality in her childhood to pass on a song that remained so vivid for Sloane throughout her life. To learn this lengthy seven-verse song at the age of twelve would have taken dedication and I suggest that the resolve to learn it arose out of Sloane's affection and/or respect for the original singer.

The emotional conflict described in the song could well have resonated with Sloane's own complex family situation. Her mother, Sarah, left her biological father, Tom Frost, and lived with William Clegg. Sloane adopted Clegg as her surname, Clegg's name appears on her marriage certificate and that of her sister, Bertha, and she had written permission from Clegg for her marriage to John Phillip Mountford (Lowe 2003). This suggests a changing dynamic between Sloane and the two most important men in her early life; the loss of her biological father through her parents' separation and the stepfather who took responsibility for the daughter he gained through the relationship with her mother. Loss and belonging in relation to significant male figures are important themes in her life.

In the late 1940s, Sloane remarried and her second husband, Fred Sloane, forbade her mentioning the name of her ex-husband.[12] Once again, an intriguing ambiguity about the terms of separation from the original husband continues a family pattern begun by Sloane's mother, a pattern that occurs in the words of *Green Bushes*. If we accept traditional folk song performance as ritual, the connection between Sloane's life experiences and the song is evident:

> During the enactment of the ritual, time is concentrated in a heightened intensity of experience. During that time, relationships are brought into existence between the participants and the model, in metaphoric form…In this way the participants not only *learn about* those relationships but actually *experience* them in their bodies…sometimes to the point where the psychic boundary between the mundane and the supernatural world breaks down so that they leave behind their everyday identity. (Small 1998:96, italics in original)

Sloane's itinerant childhood influenced her musical repertoire of old traditional songs of European origin, Australian bush songs and popular music from the nineteenth and early twentieth centuries. Her early exposure to many different musicians and the influence of her musically talented mother gave her versatility and a breadth of repertoire to match her powerful personality. Sloane accepted the influence of the paranormal in life, as her account of the Leopard Boy attests,[13] in relation to the death of the bushranger Ben Hall and the song she sang commemorating this event. The symbolism in the traditional songs she sang is a notable feature of her repertoire.

Green Bushes is a song rich in symbolism derived from the lyrics that are remnants of earlier related songs. The simplicity of the song allows the symbols to create a link between mythology and medieval culture and the endurance of these influences in the folk stream. While it is impossible to know what images Sloane saw when she sang the song, an analysis of the words offers a guide to the power of the song that has kept it alive in the oral and written tradition for probably more than 200 years:

> The starkness and simplicity of folksongs is deceptively simple, for all extra material has been discarded; yet the powerful images are never weakened or lost…If the life symbols were not present the song would have disappeared long ago…The literal reading of the ballad plots often conceals an older and deeper flow of images which should be re-examined as a sequence of pictures, in much the same way as dream or visionary sequences appear to the inner eye. (Stewart 1977:27, 46)

In Sloane's version, remnants of past singers remain in the distinctive phrasing and her pronunciation. Only one word (verse 7: 'false heart-*id*') is clearly identifiable as a reproduction of Archer's cockney accent, and this sound in translation represents the powerful personal relationship that placed the song in Sloane's repertoire. The influences from other singers in the lineage of the song remain as anonymous ghostly imprints.

In the four field recordings identified above, Sloane's performance varies each time with phrasing, word choice and changes from a four-verse version to a six-verse and a seven-verse version transcribed in Meredith and Anderson's book *Folksongs of Australia and the Men and Women Who Sang Them* (1979:173). The comparative process is hampered by the fact that tape speeds are variable because of the age of the original equipment and the fact that user copies are on cassette, which results in speed variations between tape players used for listening. I believe, however, that Sloane's pitch remains relatively consistent in each performance, with E above middle C as the tonic. This takes into account the speed problems of Meredith's tape recorder in the 1956 version, which gives a slightly lower pitch for the starting note. The following comparison describes the differences between the versions of *Green Bushes* using my transcription of the words reproduced above as reference.

- TRC 2539/5, 1953: verse 1, verse 3 as verse 2, verse 3, and verse 7 as verse 4. This version has a more regimented rhythm and has less freedom in the delivery.
- TRC 4/19, the 1956 version, is an outstanding performance with the verses flowing from 1 to 6 in sequence. The phrasing is longer and the vocal line sustained. There are kitchen noises at the beginning: a faint slosh of dishwater and then the groan of a drawer closing, shuffling movement and the way the voice volume waxes and wanes indicates head movement until a comfortable posture is found and the song gathers momentum. There is an audible drone, possibly the tape machine motor, but it sounds more like a refrigerator to me. It provides a useful harmonic reference point. There is a feeling in this performance that in the course of the song the singer leaves the domestic realm and enters the ballad world created by the narrative.
- TRC 2539/79, 1957: verse 1, verse 3 as verse 2, verse 3, and verse 7 is verse 4. Sloane's foot is audible tapping as a pulse or tactus, speeding up and slowing with her delivery of the narrative.
- TRC 4/13, c1960: there is a clock ticking in this version at about double the speed of the pulse or tactus in the song. Once again, the foot tapping is audible and the hint of pages rustling in a slight moment of hesitation between verses 2 and 3. Was she reading from a version that she had written out?

The different versions demonstrate two different styles of performance: two ways of translating the original narrative. In the shorter versions, Sloane has more detachment from the lyrics and the longer song sounds as if she is totally immersed in the narrative. This is typical of folk music as music that takes on a different shape and style to meet the requirements of the situation in which it is sung. The four-verse version captures the essence of the plot with a balanced structure of two verses for introduction and proposal and two verses for the old lover to discover the woman's change of heart, condemn her decision and move on. It is in the line 'no false-hearted young girl shall serve me so anymore' that we hear Archer's accent mirrored in Sloane's pronunciation of 'heart-*id*' (refer to Audio 4.2). The six-verse version has a different message. The woman has three verses in which to consider the proposal and reject the old love, the old love has one verse to lament and the woman has the final two verses to clarify that she doesn't want material gifts; she wants immediate companionship and an end to 'loitering'.

Sloane sang a wide range of songs as well as the introspective songs such as *Green Bushes*: 'treason songs' about Ben Hall and Ned Kelly (so-named because the police would not tolerate references to these heroes that might incite the masses to protest), bush songs such as the *Flash Country Shearer*, or *The Springtime it Brings on the Shearing* as it is also known, and the old ballads she had learned from her mother and grandmother.

Part three

My interpretation of the song *Green Bushes* (Audio 4.3) is part of a CD recording project conducted in collaboration with musician and sound artist Ian Blake, which seeks to translate the essence of Australian women's cultural identity captured in field recordings of the 1950s and early nineteenth-century transcriptions while creating music that adds something new to the continuing tradition. The recording, titled *Cantara*, was made in the spirit of the research advocated by Leslie Shepherd:

> It is true that informed and accurate scholarship is essential in ballad study, but it must be brought to life by actual experience of the folk tradition. More can be learnt from listening to a traditional ballad singer than by studying texts, and active participation in singing and dancing adds a dimension to academic study. Furthermore, some philosophical and metaphysical background is essential in a field that constantly reflects the changing beliefs of past generations. If we can learn to experience what lies at the heart of the ballad we shall resolve many of the questions that harass academic study. (Shepherd 1962:36–7)

While this research dates from the 1960s, the proof of Shepherd's assertion is in the personal investigation of his technique and I have certainly learned more about *Green Bushes* and the other songs sourced from field recordings by learning it to perform than I could ever have appreciated by a detached examination of the field recordings and analysis of the texts. This investigative technique offers a key to at least partially unlocking the identity of the song 'in the mind of the singer or a number of singers' (Porter 1976:11) based on a 'conceptual performance model' (Porter 1976) of Sloane's relationship with *Green Bushes* described in part two of this chapter. Using unaccompanied field recordings as evidence and inspiration, I recorded a selection of songs sung in Australia[14] using the vocal melody to determine the arrangements, abandoning the instrumentation associated with a bush band ensemble or the commonly used folk guitar accompaniment that imposed a rhythmic structure on the melodies. An extension of Shepherd's directive about active learning of ballads was my decision to listen critically to the background sounds in field recordings and treat them as an integral part of the information to be gleaned from listening to archival tapes and singing the songs. The sound world created by domestic noises and captured on the field tapes is an essential part of my research. These background sounds provided important clues about the domestic world that shaped the way in which women used and use their music. These sounds have been translated into the arrangements of selected folk songs to take on symbolic meanings. For example, the drone of insects is often heard in the background of field recordings. The swarm of bees heard in the segue between tracks 1 and 2, *A Bhanarach dhonn a Chruidh* and *As Sylvie was Walking*, signifies the transition of these women's songs from the Old World to the New. In the English and Celtic folklore of the women who brought this music to Australia, bees represent messengers and the bee swarm has a very Australian sound evocative of hot summers in which loud insect drones accompany most activities. Technical effects support the symbolic nature of the sound art and capture something of my era, a record of the 'state of the art' (of sound-art technology) to pass on through electronic sounds in translation.

The domestic sounds of dishes, clocks chiming, flies droning and teacups clinking communally are common accompaniments to the music recorded on field recordings. These noises inspired me to recreate sounds from my own memories of my mother's activities at home—the sound of pinking shears cutting out fabric on a dining table was one example—to bring my performances of the songs to life as part of my domestic world. Careful listening also provided clues about how women pitched the starting notes of their songs from familiarity with the resonance of the room and how ambient domestic noise provided a musical accompaniment for songs.

I spent many hours in my own kitchen with a digital audio tape (DAT) machine and an extremely sensitive microphone recording the noises that made that particular sound world and listening back to them. Water simmering in saucepans, the click of the stove thermostat, the sound of scrubbing vegetables, the drone of the fridge, water pouring out of a down pipe after rain and birdsong through the kitchen window are samples of these sounds. As my familiarity with the ambient sounds of my kitchen increased, so did my ability to accurately pitch the starting note by ear for songs I was learning from the field recordings when I was in this environment. Singing these songs blended with the daily rituals of cooking and cleaning. It became clear that the oral tradition was partly about learning the songs or the tunes by ear and also partly about relating them to sound reference points in one's own environment and time and this knowledge informed my translation of *Green Bushes* into a new recording.

My version of *Green Bushes* was learned from Sloane's singing on the field recordings discussed earlier in this chapter.[15] I chose this song above others in her repertoire because it had the feel of a very old song and because when I heard her version, I heard that the song was 'singing Sally'. So familiar is she with the words that singer and song have the one meaning, the rhythmic pulse of her performance matches a resting heartbeat, communicating a contemplative mood. I am intrigued by the ambiguity of the narrative and the added dimensions provided by the symbolism. The 'choir' that introduces my version of the song is there partly to acknowledge the anonymous influence of past singers on the song and partly influenced by my interpretation of the symbolism in the song: the nightingale as the embodiment of the spirit of a loved one and the colour green representing supernatural presences.[16] Having my vocal performance translated through contemporary digital technology into a wax cylinder recording highlights the transitory nature of any performance. For the duration of *Green Bushes*, whether it is Grainger's *Passacaglia* or a live performance, there is an interaction between the old version and the new, the past and the present and implicit in this is the transition of the song into the future.

The introduction to *Green Bushes* is constructed from melodic fragments treated in various ways: the song begins with a fragment of the melody time stretched in Digital Performer, the process repeated several times so that artefacts of the processing generate a warbling drone, combined with another melodic fragment that has been subjected to 'Thonk', a deliberately uncontrollable granular synthesis program (Email communication with Ian Blake, 15 April 2006). The idea was to convey through sound art the notion that a folk song performance coalesced out of fragments of many other songs sung by many other voices at different times—past present and future: indeed, through the action of sounds in translation. Following this fragmented introduction, more disciplined voices emerge as another layer of electronic manipulation treats a small part of the melody with a stereo 'ping pong' delay leading to a 'choral' layer constructed from the first line reversed with pitch and format shifted to a fourth and one octave below the original. These are edited in detail to create the suggestion of an unknown but somewhat Gaelic-sounding language, placed in a highly reverberant space (Blake 2006). Again, the desired sound image was that of vanished voices from singers past, with a hint of Irish influence as documented in the sources for *Green Bushes*, all converging as the song reassembled itself in the present. The original performance of the song proceeds: the warbling drone and delay layer continues, fading slowly.

> The song settles into an apparently straightforward unaccompanied ballad performance, but a band pass filter is being slowly applied along with samples of record surface noise, electrical hum and the odd scratch. There is a slow transition to the sound of a wax cylinder recording, fading into scratches and static, ending with the merest hint of granular ghosts. (Email communication with Ian Blake, 15 April 2006)

In the treatment of *Green Bushes*, my interpretation of what Grainger described as the song's 'unbroken…keeping-on-ness' (Lewis 1991) explores the connection between the past and the present, by choosing an accompaniment created by using repeated vocal samples of the original performance as texture and rhythm. I am also exploring the idea of the supernatural inhabiting the song, represented by fractured voices. The vocal line echoes Sloane's phrasing, her nuances and her remnant of Archer's English accent. Sloane's distinctive upward vocal portamento to the top E an octave above middle C and the downward swoop a fourth to the B below accentuates the inner core couplet of each verse leading to the melodic and textual resolution in the last line of the stanza, and this I have replicated in my performance. I have maintained Sloane's pronunciation of 'there' in the line 'But when he came there and found she was gone', because it emphasises, rhythmically, the critical moment, which almost sounds—'the-re'—like a wrinkle in time when the original lover arrives at the green bushes. It was also important to reproduce Archer's 'false heart-*id* young girl', adding a further dimension to the process of transmission: my voice translating Sloane's, as she translated Archer's, who translated his source singer, who were all translating the narrative of the lover's triangle told and retold in the unending cycle of *Green Bushes*.

Conclusion

My version of *Green Bushes* acknowledges the lineage of the oral tradition by including explicit references to the version of *Green Bushes* I learned from Sloane. For reasons that have been suggested above, the song was important to Sloane to learn and to continue singing throughout her life. Sloane's version appeals to me because of the personal links to significant people and because it offers audible links to the older variants through Jack Archer, from whom she learned the song. Porter's conceptual performance model offers a framework for investigating the interaction of *Green Bushes* with Sloane's life experiences and facilitating understanding 'of the musical process in formation, from the inside as well as the outside' (Porter 1975:9). This is the sense of 'keeping-on-ness' that I have tried to capture in my recording through the use of sound art and a performance that reflects the nuances of Sloane's unaccompanied version.

Bibliography

Balough, Teresa (ed.) 1982, *A Musical Genius from Australia: Selected writings by and about Percy Grainger*, Department of Music, University of Western Australia, Nedlands.

Blake, Ian 2006, Personal conversations during artistic collaboration on Sound Art for the *Cantara* CD.

Davey, Gwenda Beed and Seal, Graham 1993, *The Oxford Companion to Australian Folklore*, Oxford University Press, Melbourne.

Garner, Alan 1976, *The Owl Service*, Armada, London.

Gower, Herschel 1968, 'Jeannie Robertson: portrait of a traditional singer', *Journal of Scottish Studies*, vol. 12, pp. 113–26.

Grainger, P. 1908, 'Collecting with the phonograph', *Journal of the Folk Song Society*, vol. 12, no. 111.

Kennedy, Peter 1975, *Folksongs of Britain and Ireland*, Cassell, London.

Lewis, Thomas P. 1991, *A Source Guide to the Music of Percy Grainger*, viewed 18 December 2008, <http://www.percygrainger.org/prognot3.htm>

Lowe, Valda 2003, *Who is Sally Sloane?*, viewed 18 December 2008, <http://simplyaustralia.net/issue7/sally6.html>

Macdonald, Barry 1996, 'The idea of tradition examined in the light of two Australian musical studies', *ICTM Yearbook for Traditional Music*, vol. 28.

Meredith, John and Anderson, Hugh 1979, *Folksongs of Australia and the Men and Women Who Sang Them*, Ure Smith, Dee Why West.

Muir, Willa 1965, *Living with Ballads*, The Hogarth Press, London.

Occultopaedia web site, viewed 18 December 2008, <http://www.occultopedia.com/l/language_of_the_birds.htm>

Porter, James 1976, 'Jeannie Robertson's "My Son David": a conceptual performance model', *Journal of American Folklore*, vol. 89, no. 351, p. 9.

Slattery, Thomas C. 1974, *Percy Grainger, The Inveterate Innovator*, The Instrumentalist Company, Evanstown, Illinois.

Small, Christopher 1998, *Musicking, The Meanings of Performing and Listening*, Wesleyan University Press, Middletown.

Stewart, R. J. 1977, *Where is St George? Pagan Imagery in English Folksong*, Moonraker, Bradford-on-Avon.

University of Phoenix, *Live and Learn*, viewed 18 December 2008, <http://altreligion.about.com/library/glossary/bldefgreenlanguage.htm>

Unto Brigg Fair web site, *Unto Brigg Fair—Joseph Taylor and Other Traditional Lincolnshire Singers Recorded in 1908 by Percy Grainger* (1972), [With notes produced from the original album, now unpublished], Produced by Bob Thomson, Bill Leader and Dave Bland, Leader LEA4050, viewed 18 December 2008, <http://www.informatik.uni-hamburg.de/~zierke/joseph.taylor/>

Waters, Edgar 1957, *Australian Traditional Singers and Musicians*, Booklet in Wattle Record Archive Series, no. 1, Sydney.

Wimberly, Lowry Charles 1965, *Folklore in the English and Scottish Ballads*, Dover, New York.

Sound recordings

Gall, Jenny, *Cantara*, Elidor Records, 2006.

Meredith, John 1953, National Library of Australia Field Recordings, NLA TRC 2539/5.

Meredith, John 1956, National Library of Australia Field Recordings, NLA TRC 4/19.

Meredith, John c1960, National Library of Australia Field Recordings, NLA TRC 4/13-14.

O'Sullivan, Cathie, *Artesian Waters*, Larrikin Records, 1980.

Porter, James 1976, 'Jeannie Robertson's "My Son David": A Conceptual Performance Model', *Journal of American Folklore*, 89: 351 (January-March).

Shepherd, Leslie 1962, *The Broadside Ballad: A Study in Origins and Meaning*, Herbert Jenkins, London.

Stewart, R.J. 1977, *Where is St George*: *Pagan Imagery in English Folksong*, Moonraker, Bradford-on-Avon.

Waters, Edgar c1957, National Library of Australia Field Recordings, NLA TRC 2539/79.

Unto Brigg Fair—Joseph Taylor and other Traditional Lincolnshire Singers Recorded in 1908 by Percy Grainger , Produced by Bob Thomson, Bill Leader and Dave Bland, Leader LEA4050, 1972.

Endnotes

[1] This is my transcription from the field recordings.

[2] This is not a far-fetched possibility for Sloane to have sung 'silken walls', as wealthy people did in fact cover their walls with silk rather than wallpaper or paint.

[3] Grainger wrote, 'I firmly believe that music will some day become a "universal language". But it will not become so as long as our musical vision is limited to the output of four European countries between 1700 and 1900. The first step in the right direction is to view the music of all peoples and periods

without prejudice of any kind, and to strive to put all the world's known and available best music into circulation. Only then shall we be justified in calling music a "universal language".' Balough (1982:17)

[4] 'Ritual, then, is a means by which we experience our proper relation with the pattern which connects, the great pattern of mind. We cannot, of course, know the pattern which connects any more "objectively" than we can know anything outside ourselves. Knowledge of the pattern, as of everything else, is a relation between our inner mental processes and what is outside us' (Small 1998:130).

[5] Porter has developed a conceptual performance model based on analysis of Jeannie Robertson's singing of the ballad *My Son David*. He analysed nine recordings. 'Distinctive features became apparent at once, features which seemed on rehearings to recur as major interdependent wholes. The nine recordings could be grouped into a three-stage diachronic model…The structural element which does emerge as fluid during this period is a textual one' (Porter 1976:12–13). This fact is true for Sloane's singing as well. Sloane's recordings of *Green Bushes* were all made in her home, while Jeannie Robertson was recorded singing *My Son David* in her home and in concert situations. This fact and the importance placed on the background sounds in Sloane's field recordings are the essential differences in my analysis of her singing compared with Porter's methodology.

[6] In *The Owl Service*, Alan Garner (1976) explores the way in which the power of the legend of Math, son of Mathonwy, Gronw Bebyr, Gwydion and Bloddeuedd, creates an unending cycle in the Welsh valley where the story belongs. The story must be played out again and again through the contemporary lives of those who match the roles of the central characters in the tale.

[7] In Meredith and Anderson (1979), this is transcribed as 'where *she* thinks to meet me', but Sloane sings *he*. Both pronouns are found in the printed sources, representing different interpretations for different singers.

[8] Stewart describes the cyclical myth, which could underlie *Green Bushes*: 'the folk theme has remained constant to the elements of the act, which derives from the worship of a Mother-Lover-Goddess in whose control all life, all death were held.' *Green Bushes* is linked, like *The Two Brothers* discussed by Stewart, 'with the early ritual practice of the Sacred King and his Tanist Brother and Successor…at a certain time of the year, one chosen man superseded the present "king" usually by killing him. The victim represented the light part or waking year, his successor the dark season, or waning year. The story never ends, for the goddess brings the light-brother back to life in the spring, at the end of the dark-brother's reign…The resolving element in this perpetual struggle is the Goddess (in *Green Bushes* this is the "Fair Damsel") who restores…Poetically she sings the song of life, and all living things respond to her music, including her chosen dead lover' (Stewart 1972:17–25).

[9] See Wimberly (1965:46) for 'the Nightingale as messenger between lovers'; an 'Otherworld Bird' (p. 160); and the soul in the shape of a nightingale (p. 50).

[10] Stewart (1972:17) also comments on this quality of the melody: 'the compelling chant form of the melody of *The Cutty Wren* [the version he quotes is identical to Leaning's version of *Green Bushes*]—with its hypnotic call and response pattern—is far superior to the text that it expresses. Thus it is possible that powerful forms of music may help in the retention through time of poetry linked to them.'

[11] The version of *Green Bushes* sung by Sloane used on the Wattle record released in 1957 was compiled from two or more 'takes' so that the full seven verses were represented (Personal email communication with Dr E. Waters, 18 June 2006).

[12] 'I was never able to get any information regarding him, or even his name. When I raised the topic, Sal would lay her forefinger to her lips and silently shake her head' (Lowe 2003).

[13] John Meredith relates Sloane's story about Ben Hall's sister, the mid-wife who delivered Sloane. Mrs Coobung Mick, whose husband betrayed Hall to the troopers, was carrying a child at the time of the bushranger's shooting '(said by some to be Ben Hall's) and when the child was born it had thirty two spots on it, and that child was exhibited throughout the length and breadth of Australia for show purposes [as the Leopard Boy]. The spots were supposed to correspond with the thirty-two bullet wounds in Ben Hall's body!' (Meredith and Anderson 1979:165; <http://www.informatik.uni-hamburg.de/~zierke/joseph.taylor/>).

[14] The songs recorded are: *A Bhanarach Dhonna Chruidh, As Sylvie was Walking, The Bonny Bunch of Roses, The Stockman's Last Bed, The Female Rambling Sailor, Reedy Lagoon* and *Green Bushes*.

[15] I have re-ordered the verses as shown below:

As I went a-walking one morning in spring,
To hear the birds whistle and the nightingale sing,
I spied a fair damsel, so sweetly sang she,

'Down by the green bushes where 'e thinks to meet me.'

'Oh what are you loitering for, my pretty maid?'
'I'm a-loitering for my true love, kind sir,' she said.
'Shall I be your true love, and will you agree
And forsake the own true love and go along with me?'

'Oh, I'll buy you fine beavers and fine silken gowns,
I'll buy you fine petticoats flounced to the ground,
If you'll prove loyal and constant and free,
And forsake your own true love and go along with me.'

'I want none of your petticoats nor your fine silken robes
I was never so foolish as to marry for clothes,
But if you'll prove loyal and constant and true
I'll forsake my own true love and go along with you.'

'Oh come let us be going, kind sir, if you please,
Come let us be going from under those trees.
For yonder is coming my true love, I see,
Down by the green bushes where he thinks to meet me.'

But when he came there and found she was gone,
He looked all around him and cried quite forlorn,
Saying, 'She's gone with another and quite forsaken me,
So adieu to those green bushes for ever,' cried he.

'I'll be like some schoolboy, I'll spend all my time in play,
For I never was so foolish as to be lured away.
No false-hearted young girl shall serve me so any more,
So adieu to those green bushes, its time to give o'er.'

[16] See Note 8 and <http://altreligion.about.com/library/glossary/bldefgreenlanguage.htm> and Stewart (1977).

5. Ancient and modern footprints: music and the mysteries of Lake Mungo

Adam Shoemaker

A moment. September 2005. As a group of us walked through a meandering sand path in outback Australia, bordered with saltbush, we crested a rise. Below was an unremarkable basin: greyish brown, sparse and treeless. The colour of the cloudy sky mirrored the hue of what lay beneath our feet. Around us were the enveloping murmur of wind and the crunch of boots on sand. There was nothing to suggest that it was in any way special; it could just as easily have been any one of the hundreds of such vistas in the Willandra Lakes area, located in the far south-western corner of the state of New South Wales about 120 kilometres north-east of the Victorian town of Mildura.

What lay in front of us, however, was exceedingly special. It was a record of human habitation more than 20 000 years old—an evocation of lived heritage, which, in its depth and breadth, was unlike any other on the face of the Earth. It was a footprint site, one that suggested the activities, the stories and the sounds of Indigenous Australia over millennia. It was also a unique place in which echoes of the past—visual and aural—were unified. Ironically, we were all silenced—almost struck dumb—by the impressiveness of a place that was itself so full of centuries of sound.

Of course, the entire Willandra Lakes World Heritage Area is unique. Together with the Great Barrier Reef and Kakadu National Park, it is one of Australia's three most significant sites of global patrimony. In 1968, the oldest Indigenous Australian skeletal remains to be discovered in mainland Australia were uncovered on the shores of Lake Mungo, in the heart of the Willandra. That female skeleton was eventually dated to an age of more than 40 000 years: 'the oldest demonstrated ritual cremation anywhere in the world' (Lawrence 2006:17).

This discovery, which colloquially became known as 'Mungo Lady', revolutionised global perceptions of the nature and longevity of Aboriginal and Torres Strait Islander life in Australia. It gave rise to various new theories of human occupation of the island continent, to the interrelationship between inland waterways and nomadic peoples and it opened up scholarly interpretations of what is now called 'palaeobiology' and 'palaeopathology'. These became sites of enormous pride for the Indigenous elders of the three tribes in the area: the Paakantyi, the Mutthi Mutthi and the Ngiyampaa. The continuing, central role of those traditional owners is one of the key elements of the entire Willandra area.

In 1981, the area was officially designated a World Heritage Area by UNESCO. The proclamation of that status was recognised in September 2006 by a commemorative festival, which featured storytelling, re-enactment and, significantly, musical performance. Despite its scientific and Indigenous significance, however, the area popularly known as Lake Mungo or just Mungo is the least known (and least visited) of the Australian World Heritage zones. More than half a million tourists travel to the Great Barrier Reef and to Kakadu National Park annually; the figure for the Willandra is approximately 10 per cent of that number, or just less than 50 000 per annum. As I will argue, however, the scientific, historical, cultural, performative and archaeological importance of the Willandra is—at this very moment—more significant than it has ever been.

For this is a truly interdisciplinary story, one that surprises, confounds expectations and that has the potential to—yet again—revolutionise our understanding of ancient history and of original sound-scapes. It is one that brings the humanities, the creative arts and the sciences together in an intellectual coalition of great strength. It is one in which both senses of the term 'fieldwork' are comprehensively explored: where experiential education and (literally) groundbreaking archaeological research are aligned in a unique way. It is a project in which the contribution to Indigenous history by Aboriginal and Torres Strait Islander people themselves is an irrepressible force. It is one of the clearest examples of the intersection between globalisation and Indigenous culture—a juncture that holds both risks and the potential for social justice as well as recognition.

Nearly four years ago, Mary Pappin, Jr—one of three Indigenous trainee rangers in the Willandra—walked over the rise that I mentioned at the outset, entered into a large pond-like depression and noticed a single, remarkable echo of the past etched into the ground: an unmistakably human footprint. That careful observation by Pappin has changed, yet again, the perspective of the Lake Mungo National Park (whose famous human skeletal remains were discovered some kilometres away). Working with the respected archaeologist Professor Stephen Webb, the Three Traditional Tribal Groups Elders Council authorised the painstaking excavation of the area. No-one could have predicted the extent, the import or the vividness of that discovery.

From one partially obscured impression, the team of Indigenous and non-Indigenous researchers uncovered, by September 2005, more than 250 footprints. By December of that year, the number had risen to 450 and, according to Webb, there are likely 600–700 prints in total in just that one area; it is, quite simply, the most significant footprint site on the planet. Expressing this another way, there are more extant footprints in the Willandra than in all known sites in the rest of the world *combined* (*Catalyst*, ABC Television, 31 August 2006). What is particularly remarkable is the incredible clarity of the record. Bioluminescence, or optically stimulated luminescence (OSL), testing has established that the tracks are truly ancient. They were formed approximately 20 000 years ago but they are so detailed and evocative that it seems one is following the walkers just minutes after they were present. One sees the unmistakable signs of mud between the toes, of children who have leapt and gambolled and changed directions suddenly to follow their mothers, of five men who were all running very quickly in the same direction.

I argue that the site is also one that is imbued and 'inspirited' with sounds—of the era 20 centuries ago and the present day. And, on the day I was introduced to it with students from our university, we examined one of the most baffling footprints I have ever viewed. There were at least 14 impressions of what everyone termed the 'one-legged man': these were right footprints only, each separated by more than one metre, with no signs of hopping or vaulting. The size of the foot was very large—size 11 or 12 in North American parlance—which scientists tell us equates to a height of at least 194 centimetres (Webb et al. 2006). There were audible gasps among the entire group when faced with this seeming mystery.

The overarching emotions shared by everyone were reverence and awe. A footprint—the most simple of human records, and one that is so evanescent it is normally washed away on the beach by the incoming tide—was here, frozen in time in a way that was incredible. It is, therefore, a human site, one that provides a glimpse of ancient history as a simulacrum. It is an Indigenous Australian site that has great spiritual significance as an echo of specifically Aboriginal ancestry. It is explicitly an artistic site in which the human design of footprints forms an unmistakeable canvas—a pattern that is as entrancing as it is creative. It is an irrepressibly musical venue, in which the acoustic environment comes to the fore in a myriad ways. Thus, faced with its spectacle, all viewers become intensely focused listeners as well. All present are transformed to become members of an audience that apprehends real and imagined sounds, past and present alike: those outside the frame of the site itself (such as the echoes of far-off thunder) and the literal echoes of the past suggested by the tracks themselves.

Finally, this is a scientific site, where the latest technique of, for example, ground-penetrating radar shows that there are 22 separate tracks and, astonishingly, that what is visible is about only 10 per cent of what could potentially be revealed. As Professor Steve Webb has stated, 'There could be seven or eight layers of footprints there laying [sic] on top of each other; eight phases of these people hunting and gathering' in a form of three-dimensional revelation (Personal communication, 4 August 2006).

All of these dimensions of the site are intertwined in a unique partnership: non-Aboriginal students of archaeology are able to visit the area only when given the permission and careful directions of Indigenous elders, while non-Indigenous pastoralists and scientists pool their knowledge, as well as their trust, in so doing. The overarching sense of being there is, however, one of respectful mystery—of sound, even within the silence. And of storytelling.

For the site is imbued with an inescapable sense of artistic narrative. The immediacy and familiarity of the record immediately lead to hypotheses, not only about the age of the area but about the type of soil—which is magnacite, a rare amalgam of magnesium and carbonate—which enabled the mud prints to be 'frozen' in time for 20 millennia. All the theories concerned imagined behaviour and the recording of a very personal sense of history. How could it be that a one-legged man could have survived 20 000 years ago? How did he lose his leg—was it burned in a fire? Was one leg bound against his body? Did he carry a walking stick for support (and why were there no indications of depressions from it in the mud)? And how could he possibly have hopped so far when no human beings alive today could have done so? Could he have been kneeling on a canoe with his left leg while propelling the watercraft with his right? How could these mysteries be solved?

Perhaps we can say that these mysteries are investigated, if not actually solved, through imagination (this being in a sense the very nature of a hypothesis). As we imagine each possible scenario, each imagined story, sound becomes an even more powerful element. Thus, the 'mind's ear' as well as the 'mind's eye', not to mention the 'mind's action' (the kinetic element), are all brought together on this ancient stage.'

What this indicates is that a particularly powerful sense of *performance* is present on, and in, the site. This is what is so striking about the Willandra Lakes discovery: that, perhaps, *three* different types of human experience meet there (in symbolic terms) on the claypan—ancient history, forensic science and the creative, musical arts (there were expressions of dance, of design, of singing, of semiotic marking). There was, in and on the soil, an overriding atmosphere of discovery and—paradoxically—of blindness. For whatever we observed in hypothetical terms, we were all acutely aware of our own limitations, of how we, as latter-day, twenty-first century observers, were probably missing most of the visible signs in front of us.

One of the clearest ways of throwing this paradox into relief is to consider the inspired leadership of the Lake Mungo traditional owners, who decided to invite three Indigenous trackers from Central Australia to visit the footprint site in 2006. The three were traditional Pintupi/Warlpiri people, all of whom had not had contact with European Australians until they were in their thirties. According to Professor Webb, who was present at the time, they were completely silent when they first saw the area; then all three began to whisper among one another. The single word they said was '*Djukurrpa*', the Pintupi-Luritja term that we in the English-language world have imperfectly translated as 'the Dreaming'. In other words, for the visitors, the footprints were a visible manifestation of continuing religious belief, of the persistence of past sacredness in the present. This was an unsurpassed example of ancient and modern stories coming together in a fascinating moment.

At the same time, the trackers observed an immense amount—far more than anyone else had noticed. They found almost invisible skid-marks, showing where the five running men had thrown spears (unsuccessfully) at their prey—likely a kangaroo. They noted a small physical record where one of the people (perhaps a child) had been doodling in the mud. They were totally unperturbed by the one-legged man and found evidence of small depressions where he had in fact been leaning on a pole while he hopped. They showed where he stopped, where he threw a stick, where he jumped. In the words of Professor Webb, 'these trackers were not at all fazed by the footprints; for them it was simple proof of the fact that if you did not adapt you would die' (Personal communication, 4 August 2006).

Moreover, these insights—the reverential awe of the trackers when first faced with the site, as well as their deep apprehension of its many-layered significance—raise the whole issue of comparative, cross-cultural understanding. Equally, it proves just how indispensable the specialised Indigenous contribution to Australian research can be, not just at Lake Mungo, but wherever human contact has taken place on this continent. At the same time, the depth of the Mungo discovery suggests the profundity of Aboriginal music (and lyrical poetry). Put simply, there is much to sing about at the footprint site.

Consider this from a slightly different perspective: our discovery of a unique landscape such as this one inevitably attunes the senses in a different way. The initial human reaction is one of silence and stillness: the view of the ancient world on display literally takes speech away. A key realisation, however, is that, while speech is silenced, one's hearing is not. Indeed, *all* the senses are heightened; the visual impressiveness of the array of footsteps makes one look even more closely. One wants to touch the soil, to feel it under bare feet, to try to connect in a kinetic way with the human map of behaviour that has unfolded.

Then, a different form of apprehension takes over. One truly listens in a new way to ambient and imagined sounds: to the wind whistling over the rise; to the echoes of bird calls; to the sounds that our own footprints make on the calciferous surface; even to one's own inspiration (in all senses of that word).

It is at that precise moment that sounds become translated and transported in striking fashion. And, not surprisingly, this is a moment of blinding inspiration and of revealing insight. What *is* surprising is that one's individual senses do not suffer as a result of others coming into play: visual and sonic clues act in tandem and are mutually reinforcing.

In this respect, no written essay can do justice to the experience. What a chapter such as this one *can* do, however, is to signal the fact that the musicological debate about the 'predominance of vision over audition' does not apply in this case. Many theorists have agreed with Murray Schafer, who argued in his landmark 1977 text *The Tuning of the World* that, '[i]n Western culture truth is not "heard": (as in Jewish religion) but "seen" and that this "contributes to a lack of sonological competence (competence for understanding sound formations) in our culture"' (quoted in Laske 1978:395). In this case, however, where the simple complexity of the Lake Mungo footprint site enriches all of the senses simultaneously, sight and sound do not vie for ascendance. Rather, there is a form of impressive harmony produced by the depth and uniqueness of the human record that surrounds any viewer, any listener, any observer, any hearer of that special environment.

Another way to express this is to say that it is nearly impossible for any human being not to react to the site in this way. It has an inescapable, inbuilt narrative or thrust towards storytelling. It is replete with latent mystery. And, in terms of what has been called 'acoustic ecology' (or the 'study of the effects of the acoustic environment or soundscape on the physical responses or behavioural characteristics of creatures living within it'; Schafer 1977:2), Lake Mungo is like no other Australian landscape. Even more: as small, coloured flags were painstakingly placed next to each footprint by archaeologists and their collaborators to mark the trajectory of each imprint on the soil, the overall visual frame altered again—to produce a *different* sonic record.

Much as Indigenous sand and bark paintings often portray an inbuilt 'eagle's-eye view' of the landscape, as if the artist were adopting an aerial perspective of special sections of country, the flagged track-ways at Mungo looked, and felt, for all the world like the notations of a musical score. Semiotically, then, the land signifies music and is translated into a musical pattern. Fascinatingly, this applies not just on one plane but on many, so the score is both on the surface and subterranean. This evokes Stephen Muecke's (2004:106–17) powerful observation about the 'subterranean river of blood' in Arrernte culture; proof of the fact that 'the Law' (and evocations of suffering) in Indigenous Australian society are deep matters indeed.

If this is not fascinating enough, there is more—much more. For the location suggests not just current sounds, or the shape and feel of a composition, but ancient sounds of 20 000–23 000 years ago—and it does all of this at once. Seeing the extant prints of a kangaroo that hopped centuries ago immediately conjures up the image of the creature and the sound of its distinctive thumping echo on the ground. Seeing the crisscrossing tracks of children playing brings to life the ambient noise of shrieks of laughter and pleasure. Noting the mark of a spear on magnacite brings to the mind's ear the special sound of skidding wood on sand…and so on.

At every level, therefore, the Lake Mungo site operates like a three-dimensional chess game with an inbuilt sonic register. The site is undoubtedly liberated by science but it is also humanised by musical sounds in translation. It is a performed place in terms of ritual, religion and storytelling, but also via lyricism and music. It is undoubtedly narrational and has intrinsic performativity and it is also—at one and the same time—a natural sound-scape.

In all of these multiple ways, the Lake Mungo discovery proves the appositeness of what Steven Feld (1984) has termed 'qualitative and intensive comparative sociomusicology'. Put another way, particular venues such as Mungo, which are full of sonic and visual meanings, can be thrown into comparative relief to illustrate a whole range of translational conclusions. As Feld (1984:385) continues:

> Comparative sociomusicology should take the tough questions and sort them out with the best materials available for detailed comparison: the thorough, long-term historically and ethnographically situated case study. The meaningful comparisons are going to be the ones between the most radically contextualized case examples, and not between decontextualized trait lists.

I agree with him, explicitly and implicitly. I believe that, like a classic Shakespearian play, Lake Mungo provides the stage for imagined and ambient sounds. The surprising aspect is that both can be apprehended simultaneously, and in harmony.

Given all of this significance, it is no wonder that the footprint site—though still relatively unknown—is rapidly changing the face of ancient Australasian history. The discovery featured in the June 2006 issue of *National Geographic*, a television treatment focusing on the site was aired on the Australian Broadcasting Corporation's science show *Catalyst* on 31 August 2006, and the Central Australian Aboriginal Media Association (or CAAMA) accompanied the three Central Australian trackers on their journey and was preparing its own documentary treatment of the encounter. One common feature of all of these approaches is that they underline the uniqueness of the Willandra and the incredibly local, specific and grounded nature of the site. It is a visible manifestation of traditional Aboriginal law, in which the real power resides in and beneath the soil in a place of spiritual, scientific, artistic and cultural depth.

At that very instant, however, in the same few moments that one can read the previous sentence, one realises that the footprint site is also simultaneously a world issue, an artistic challenge and a manifestation of globalisation. This is also, I would argue, a key marker of the contemporary Aboriginal and Torres Strait Islander universe. In this way, the most prominent, the most local and the most grounded signifiers of culture in Australia are *simultaneously* the most international, the most migratory and the most emblematic of that country's images globally—as relevant in Toronto as they are in Taipei.

Just as the images in the Indigenous Australian section of the recently opened *Musée de Quai Branly* in Paris conjoin the past and the present in a moment of instantaneous transference, so the footprint site represents a unique coalition between science and the new creative arts. How so? Because every dimension of the external treatment of the footprint site is one of globalism: images of the area are readily downloadable from at least 10 different locations on the Internet (including *National Geographic Online*) and the importance and putative authenticity of the site were first validated via a major article by Webb and Cupper in the *Journal of Human Evolution* (accepted and published electronically on 26 October 2005).

Even more: the desires of the Three Traditional Tribal Groups to slowly and steadily control the release of imagery from the footprint site (itself an attempt to assert a measure of Indigenous sovereignty over the discovery) were thwarted because the *Journal of Human Evolution* posted the relevant article on its web site many weeks before the hardcopy of the journal was released. This led to the inadvertent leaking of the news of the footprint findings more than a month earlier than the elders had planned, forcing them to orchestrate a rapidly organised media release about the track-way just before Christmas 2005.

Therefore, the same communication tools that the elders employed for their media campaign had already robbed them of the means to announce the footprint discovery in a strategic, Aboriginally controlled fashion. It also markedly lessened their ability to exercise the same degree of custodianship over the intellectual property produced by, and inherent in, the site—in particular, the world-wide dissemination of the dramatic colour photographs taken of the footprints. Therefore, all of those factors—unexpected early news of the discovery; the release of (unauthorised) images; the professional trajectory of the international archaeological community—participate in the same three conundrums. These are:

- that indigeneity is often taken as a proxy for timelessness but contemporary time is very much of the essence
- that indigeneity is frequently understood as being incredibly specific to certain landscapes and local cultures, but its validation is always according to putative international standards and categories (whether they be the

'dating' of deposits, an assessment of the assumed authenticity of records or the dissemination of knowledge)
* that indigeneity is mistakenly thought of as being (like the footprints themselves) solely ancient while its involvement with modern technology is vital, irresistible and subject to continuing change as a condition of existence.

Thus, contemporary science meets the Indigenous arts, aligned in a new moment in which each liberates the other. That said, the Lake Mungo footprint site shows so clearly that while technology cannot be realistically ignored it must be placed in ethical context on each occasion, in each visit, in each article that is written. As ethnomusicologist René Lysloff has written:

> Technology privileges researchers, distancing them from the object of research—whether musical or human—and allowing them to control it. Indeed, the sound object becomes a true object: isolated from the noisy chaos of real life in the field it becomes analysable, frameable, manipulable [sic], and ultimately…exploitable. (1997:207)

What clearer elucidation of the challenges of translating sounds (and images) can there be?

I want to emphasise that there is not a simplistic or ready-made formula about the value and ease of maintaining ancient cultural heritage in the modern world. I am, however, convinced that an interdisciplinary approach—or a trans-disciplinary adventure between the sciences and the creative arts—provides unique, and more exciting, strategies for teaching, learning and understanding. At the same time, developments in the new media arts enable us to visualise sensitive, friable, vulnerable, human artworks (be they skeletal remains, tapestries or, as in the present example, footprints) in a manner that enhances our understanding of human creativity. The visualisation afforded by, for example, three-dimensional animation and immersive sound can make the experience of touching history as vivid as touching the original exhibit, whether or not that is physically possible.

Similarly, the conservation needs of, say, rare tapestries imply a level of reduced light that makes viewing them nearly impossible for the sight-impaired. Happily, the digitised and dramatised version of that tapestry on the interactive computer screen brings the artefact to life as never before. Instead of simply 'being there', we *become* there—and this concept of formative discovery is a fascinating and liberating one for those who teach art, design, sculpture, new media arts, music, art theory and many related disciplines. Let me give a further instance of the context of the footprint site. It is that every discovery carries with it both risks and triumphs. In the case of the Willandra Lakes site, the very process of opening up this vast treasure chest of knowledge has made the site both more secure *and* more vulnerable—at exactly the same moment.

It is theoretically more secure because the Australian Museum in Sydney has organised the creation of plaster casts of some of the iconic prints and has arranged ground-penetrating radar scanning of the location, 80 per cent of which has been completed. Those very processes (and others associated with them) have, however, put the site at increased risk. The very opening up of what has been hidden for more than 20 000 years has meant that the elements can now attack the site, feral animals (and humans) can potentially degrade the landscape, scientists (instead of Aboriginal people) can claim the discovery—and have done so—and, as I have observed, the control of imagery and sound (in intellectual property terms) has become very difficult to maintain.

So, while there has been an intense flurry of concentrated activity, to date, the traditional owners of the area—those who discovered the site, who have a custodial relationship with it and who, more than anyone else, feel its sacredness—have up to now had no success securing support for a research, art and education centre that could interpret and explain the footprint track-way. Nor have they garnered enough backing to create a 'keeping place' or archive of remembrance, for this, and future, discoveries and repatriations. What they *do* have are very clear aspirations for its future control, use and management—and to imagine a future that is radically translated in their favour.

This could come, and many wish it so, but the future prospects for the site are both liberating *and* limiting. The Australian Museum and the NSW Parks and Wildlife Service (NSWPWS) are engaged in establishing a so-called 'management plan' for the zone, but serious questions remain. Above all, in a World Heritage Area such as the Willandra Lakes, *whose* world is it, and whose heritage is being managed? Australia has a stake here, as do instrumentalities such as the NSWPWS and governments. Equally, however, the United Nations has a role to play, via UNESCO—as does the international and national archaeological community and, I would argue, those of us in academia (especially in music and the creative arts) who recognise the pivotal importance of such discoveries for our own pedagogy. The pastoralists of the region also play a central part: they are involved in the joint management of the park at all levels. Crucially, so are the Indigenous traditional owners. Their foundational roles of discovery, pride and custodianship underlie the entire future, and the future performance, of the Willandra Lakes World Heritage Area.

Meanwhile, the ultimate irony is that the site itself has essentially disappeared at present; it has been effectively reburied with sandbags for its own protection. This is because its uncovering in 2005 led to serious damage in a freak weather event in June 2006, when three consecutive days of rain filled the area with water, to be followed by several days of sub-zero temperatures and frost. Ice penetrated the aeons-old striations and caused many to be damaged. When temperatures rose, the ice expanded and melted. It is sobering to think that in just 20 months, human intervention occasioned more damage to the footprint site than 20 000 years of weathering and erosion.

This raises some powerful practical and ethical questions: does the 'translation' of a special area such as this one into contemporary understanding inevitably imply its degradation? Who really owns the site? Should the footprint track-way therefore remain covered? In ethical terms, should it be allowed to erode, since years of natural weathering allowed it to become uncovered in the first place? Should a fully fledged replica be built, either above or near it—or would this be a travesty of any notion of authenticity? Should the site be roofed? Should we continue to visit the area for educational purposes? Should it be dug up and relocated? Should it be installed in a university, an art gallery or a museum? Should it *become* an open-air museum in its own right? Should it become established as a focal point for Indigenous-owned cultural tourism? Should it not be visited at all? And who will make these decisions?

The point is that these are potent issues of heritage, history and human behaviour, and each decision has an implication, a value and a narrative of—and through—the authenticity that surrounds it. Ironically, given the ambivalent relationship that some of the elders have had with the communications media, the very revelation of the footprint site in face-to-face mode has made it even more essential that the future preservation of the site is, for example, through the new media arts, via three-dimensional visualisation and animated imagery. A version of the Willandra track-way that utilises the sophisticated artistic and imaging techniques developed for such productions as *Walking with Dinosaurs* could just be the answer, and will liberate—rather than constrict—the storytelling potential that lies there. This is a significant area in which the academy (and within this the creative arts) can play an active and pivotal role—and it will.

There is no doubt that the Lake Mungo footprint site is an unparalleled case study, in which questions of philosophy, translation and authenticity take centre-stage—as do issues of the control of imagery, tourism, sound-scapes, the arts, archaeological discovery and the redefinition of Australian (and world) history. Finally, there is the question of governmental clarity: a World Heritage Area is the very clear responsibility of the Australian Federal Government while, in the main, the Lake Mungo area has been managed in the past 25 years at the local, conservational and developmental level by the State Government of New South Wales. The relationship between the two is one crucial index of the harmony (or lack of partnership) that characterises the management of the Willandra Lakes zone.

One can only hope that the narrative and sound-scape so deeply imbued in Lake Mungo continue to be evoked in such a way that the Aboriginal owners, discoverers and true interpreters of the site continue to act as its creative custodians. The question of who writes that future into being—of who literally 'authors that authenticity' and, perhaps, who performs it in translation—is both searing and pressing. And that provides, in world terms, the greatest challenge of all.

References

Feld, Steven 1984, 'Sound structure as social structure', *Ethnomusicology*, vol. 28, no. 3.

Laske, O. 1978, 'Review of *The Tuning of the World*, by R. Murray Schafer', *The Musical Quarterly*, vol. 64, no. 3.

Lawrence, Helen (ed.) 2006, *Mungo Over Millennia: The Willandra landscape and its people*, Maygog Publishing, Sorell, Tasmania.

Lysloff, René T. A. 1997, 'Mozart in mirrorshades: ethnomusicology, technology, and the politics of representation', *Ethnomusicology*, vol. 41, no. 2.

Muecke, Stephen 2004, *Ancient and Modern: Time, culture and indigenous philosophy*, UNSW Press, Sydney.

Schafer, R. Murray 1977, *The Tuning of the World*, McClelland and Stewart, Toronto.

Webb, Steve, Cupper, Matthew and Robins, Richard 2006, 'Pleistocene human footprints from the Willandra Lakes, southeastern Australia', *Journal of Human Evolution*, vol. 50, no. 4.

6. Translating the *shigu* from the streets to the stage

Amy Chan

Video footage relating to this chapter is available online at:
http://epress.anu.edu.au/titles/sounds_translation_citation.html

The large barrel-like drum the *shigu* is part of a percussion ensemble that accompanies the Chinese lion dance during New Year celebrations and other folk festivals. The din created by the *shigu*, together with the gong and cymbals, heralds the arrival of the lion-dance troupe and also serves to symbolically drive away evil spirits. This chapter seeks to chart the abstraction of this drum from its traditional context and function to its subsequent appearance on stage. It attempts to throw light on the changes in sound when 'translated' from one domain to another and raises the question of how this transplantation has impacted on the ability of the *shigu* to represent Chinese culture in Malaysia.

Before I proceed, let us take a closer look at the contextual background from which the *shigu* emerges. As mentioned, the large drum is part of the accompanying ensemble of the Chinese lion dance. The dance is a celebration and practice steeped in ritualism, martial arts and clan rivalries. As part of the ritual, the noise-making instruments serve to herald the arrival of the 'lion' and to exorcise the performance place of evil spirits. The performance is usually held as a street procession leading to the entrance of a shop or house whose owner had requested the performance (Slovenz 1987). This part of the ritual is complemented with blessing and good fortune, as symbolised by the snatching of a red envelope containing money and a bunch of green leaves (usually lettuce). There are two versions of the lion dance—a northern and a southern—and its performance during festivities is practised in China and Chinese diasporic communities (Slovenz 1987; Sarwar-Yousof 1986; Johnson 2005a, 2005b). Two performers hidden under the papier-mâché head and its silk tail animate the 'lion'. The performers manipulate the head and tail in vigorous expert martial art movements to the drumming of the *shigu*. Very little research has been carried out on the sound aspect of this performance in its current practice. Much of the current research focuses on the lion-dance performance and its ritual implications. Slovenz, however, did note in her account the change in drumbeats of the *shigu* that seemed to signal the change in movements of the lion. Thus, the drum plays an important role as a communicator—to the general public of the entrance of the lion to the scene and to the lion-dance performers of change.

Politicising the lion dance in Malaysia

In Malaysia, it is the southern version of the lion dance that is most widely performed (Sarwar-Yousof 1986). It has the same ritual significance as elsewhere, however, the lion-dance performance and its socio-cultural significance have evolved over the years, involving some 'reworking and reinvention' (Carstens 2005:168). For example, although street performances are still apparent, the performance has also evolved into annual competitions held at the Genting Resorts near Kuala Lumpur, attended by national and international performance troupes.[1] Although an interesting topic, it is, however, not the focus of this chapter. As described above, the lion-dance performance holds import and significance for the Chinese community in Malaysia. The Malaysian population comprises 64 per cent ethnic Malays/Bumiputras,[2] 26 per cent ethnic Chinese, 7 per cent Indians and 3 per cent minority and mixed ethnicities (Department of Statistics 2000). The lion-dance performance is significant because the dance is representative of the Chinese community in Malaysia and their struggle for recognition and acceptance as equal members of the Malaysian political entity. According to Carstens (2005), the lion-dance performance is an important aspect of Malaysian Chinese culture because it epitomises 'the lessons learned by generations of immigrants: that success in life and good fortune did not come easily, but ultimately derived from a combination of skill, hard work, and auspicious circumstances'.

Within the Chinese diaspora, the street performance has held an honoured role of representing Chinese culture, community and tradition vis-a-vis a majority community (Slovenz 1987; Johnson 2005a). As a folk practice that dates back to the Ming and Qing Dynasties (Carstens 2005:168), it could even serve as a cultural link to a traditional past and place of origin. In Malaysia, however, this representation is eminent.

In the decades between 1970 and 1990, the lion dance took on a highly politicised position in the ethnic tension-ridden social landscape of Malaysia. The national cultural debate began with the National Culture Policy of 1971[3] that stipulated that only culture practiced by the indigenous people of Malaysia would be recognised as part of the national culture. In practice, this policy welcomed mainly traditional art forms and practices of the Malay community. The exclusion of other traditional cultures (namely those of the Chinese and Indians) was compounded when government ministers such as the Minister of Home Affairs, Ghazalie Shafie, labelled these cultures and practices (including the lion dance) 'archaic', 'extinct', 'futile and a waste of time' and a 'hinderance [sic] to the emergence of a national culture' (Kua 1990:10–11). *The Straits Times* (20 May 1979, quoted in Kua 1990) reported that Shafie said specifically of the lion dance that, 'with its musical accompaniment, [it] could not be easily accepted as a Malaysian dance form and music'. He went on to say that 'the dance had originated from China where it was now extinct and as such it could not develop further and be accepted by all'.

Representatives of the Chinese community—the Democratic Action Party (the opposition of the incumbent Coalition Party), the Chinese Malaysian Association and the Malaysian Chinese Association—reacted against this, arguing for a multicultural national cultural policy and identity that would recognise the contributions of all Malaysian citizens regardless of ethnicity. They persisted that the lion dance was an inextricable part of Chinese culture and its practice should be absorbed into a Malaysian national culture (Kua 1990; Tan 1992; Carstens 2005).[4] The fragmented Chinese clan associations (which traditionally were rivals and these rivalries were displayed during the lion-dance performance) were brought together to form the Federation of Lion Dance Associations in the early 1970s (Carstens 2005:152–3) to create a united Chinese front against the perceived adversarial charges of the pro-Malay government and its policies. It was in this context—the politicisation of national culture—that the lion dance began to take on a symbolic status as representative of Chinese struggle for recognition in Malaysia.

The ethnic tension subsided somewhat after an operation to clamp down on political dissidents, 'Operasi Lalang', in 1987 (see Crouch 1996:80–2, 106–13). In 1991, Mahathir Mohamad, the then Prime Minister of Malaysia (1982–2003), announced his new vision for a fully developed Malaysia by 2020 (<http://www.epu.my>). This plan placed emphasis on development not only economically but in education and culture. One of the first challenges to this aim, he contended, was to confront the ethnic divisions and tensions that were ingrained in the sociopolitical landscape of Malaysia. He campaigned for a 'Bangsa Malaysia', which in simple translation could mean a 'Malaysian race' or 'Malaysian nation'. This concept engendered many ensuing discussions and debates about what it meant and the practicalities of its application. Mahathir himself attempted a definition in 1995: 'Bangsa Malaysia means people who are able to identify themselves with the country, speak Bahasa Malaysia[5] and accept the Constitution' (Straits Times, 12 September 1995); and, in its realisation, Malaysians are not expected to 'give up one's culture, religion or language' (cited in Tan 2003). New and revised policies were put in place to achieve Mahathir's vision, which some argued gave rise to economic and cultural liberalisation (Loh 2002; Tan 2003).

This swing to an acceptance of diversity and difference from the top-most echelon of the national government opened up further discussion and debate about what it meant to be Malaysian. This fomented much discussion within communities of Malays and non-Malays (including Chinese and Indian communities). Local academics wrote about how *Bangsa Malaysia* might or might not differ from 'Bangsa Melayu' (the Malay race/nation) (*Business Times*, 19 August 1995; Sani 1992), local newspapers conducted forums on it (*New Straits Times*, 2000, 2002), the smaller political parties headed by ethnic Chinese such as the Parti Gerakan latched on to this for their political thrust (*New Straits Times*, 21 August 2000) and conferences were held in its honour (Fourth International Malaysian Studies conference in 2004). For the realisation of *Bangsa Malaysia*, it has to first overcome the incumbent, ethnically conflicted sociopolitical landscape and Malay/non-Malay altercations. Suffice to say here that *Bangsa Malaysia* has not yet become a reality, but is still in the process of negotiation. The Malaysian national consciousness is still an imagined community.

In this multicultural and more liberalised political environment, the lion dance again came to the fore in the debate about national culture and identity. While it still represents the Chinese in Malaysia, it is no longer a symbol of their struggle for recognition, but is harnessed to symbolise the Chinese contribution to a multicultural Malaysia (*Bangsa Malaysia*) by the State that once denounced them. At present, the State's definition of *Bangsa Malaysia* is based on a multicultural rubric, as demonstrated in *Citrawarna* ('Colours of Malaysia'), a street parade organised by the Ministry of Arts, Culture and Tourism annually since 1999. The theme of the parade emphasises cultural diversity as represented by the various performances (including *kompang*[6] ensemble and the lion dance), the variety of 'traditional' costumes and the popularised versions of folk melodies of the various cultural groups strung together into a medley. It is obvious in many respects that this is a performance, not so much for Malaysians themselves (for many Malaysians are not aware of the parade or its telecast the next day),[7] but for foreign tourists. All announcements and speeches are carried out in Malay and English, the (television) audience[8] is constantly reminded that there are 1.5 million people in China watching and it is claimed that this is to showcase the *'kehebatan Bangsa Malaysia'* (the Malaysian nation's greatness) to the world. That notwithstanding, it is still quite a show of Malaysia's recent turn to multiculturalism.

One of the performances in the parade is, of course, the lion dance, as the cultural representation of the Chinese community. While in the past, the State saw the lion dance as a threat to the enshrined national Malay culture, here it is a convenient instrument to indicate the State's 'change of heart'. It is important to note that despite this perceived change, the National Cultural Policy remains in place. The lion-dance performance no longer represents only the Chinese; it is harnessed towards the construction of a larger identity—that of multicultural Malaysia. It is used here to represent Malaysia's embrace of diversity and difference. Where in the past, the lion dance was a cultural marker of Chinese-ness necessarily stated and distinguished from a local Malay-based majority, the lion dance now retains its role as a Chinese symbol and also represents the Chinese contribution and presence in an ostensible display of 'multicultural Malaysia', as stated and distinguished by the regional and global stage of international tourism.

Against this backdrop, I will now reintroduce the *shigu*, the sonic abstraction from the lion-dance performance. This began in 1988 with a schoolteacher, Tan Fui Choong. Together with a friend, he came up with the idea of removing the drums, *shigu*, from their traditional ensemble of drums, gongs and cymbals, and their association with the lion-dance performance, and placing 24 in an ensemble performing in unison, incorporating martial arts and dance movements in its choreography. The focus is on its abstraction from the folk performance and its subsequent re-assemblage as an ensemble of *shigu*. The number, 24, has special significance and is based on the Chinese 'agricultural calendar'. The number of drums correlates with the 24 solar terms, a system that charts the movement of the sun through the course of a year from Earth's perspective. The agricultural calendar divides this movement into 24 nodes, which are used to mark agricultural seasons.[9] Tan formed the first 24-drum troupe. The performers came from Chinese high schools and various lion-dance associations. Presently, there are 50 such troupes in Malaysia and Hands Percussion, the troupe in focus in this chapter, is one of them.

The *shigu* and being Malaysian Chinese

In an interview with the local newspaper, Tan was emphatic that this new creation was 'uniquely Malaysian in concept' and such a troupe was 'reflective of the multi-racial scenario' (*Star*, 8 January 2002). This last point most likely meant that cultural performances in Malaysia included Malay and non-Malay traditions. The keenness to express a Malaysian identity (as opposed to a Chinese or Malaysian Chinese identity) is perhaps the result of what Tan Chee Beng (2001:215) terms 'localization—the process of growing up and being socialized in a local setting thus acquiring local consciousness and being influenced by local political and sociocultural forces'. Tan Fui Choong expressively states his association with the nationality and 'place' of this creation, despite the historical origin of the practice and tradition. The origins of the *shigu*, as part of the performance of the lion dance as a folk practice common to migrants from China, are not a concern to him. What is of concern is the re-creation of the drums as an ensemble of their own as founded and 'placed' (as in situated and imbricated by the socio-historical forces of the place) in Malaysia. Another interesting aspect Tan raises in this report is that these performances are not scripted.

From this 'Malaysian creation', we move on to a more contemporary modification of the 24-drum ensemble. Bernard Goh, previously a student of Tan, formed his own drum troupe, Hands Percussion, in 1997, in Kuala Lumpur with Eric Ch'ng. Respectively, they are the artistic director and administrative director of this semi-professional team of drummers. They currently have eight full-time performers, seven part-timers and 12 in training. Their web site (<http://www.hands.com.my>) delineates their vision as not only to 'preserve traditional Chinese percussion and its performing arts', but 'to make Chinese drumming more artistic and creative' by introducing 'new dimensions to theatrical drumming and [exploring] innovative permutations in contemporary percussion music'. On the one hand, the troupe is concerned with the conservation of the practice of this performing tradition; on the other, it is also their aim to extend the practice into a different level of artistry. This dual purpose raises an interesting paradox: here art and folk practice are expected to coexist in the same space, in the same instrument, in the same performance. Is not art conventionally on the opposite end of the spectrum from tradition and traditional forms? 'Tradition' by definition is about the maintenance of a specific practice with an emphasis on continuity (Layton 1992); art, and particularly modernist art, emphasises the transcendence of nature (Hegel 1975:29–30) and of society (Marcuse 1978). By implication, the phrase 'traditional Chinese percussion' could simply mean the instrument in question having roots in a traditional practice, but its current usage could differ tremendously from that original practice as the work of what Hegel (1975) terms 'human hands'.

This is evident in the set pieces for performance in Hands Percussion's repertoire: *Ritual of Drums* and *Dialogue in Skin*. The former comprises four parts: 'The Five Elements', 'Awakening', 'Reincarnation' and 'Sound Play'. The group's web site (<http://www.hands.com.my>) states that it is their aim to present the drums as more than mere 'noisy instruments', rather as 'communication tools through which thoughts, emotions and feelings can be expressed'. Composed by Goh, *Ritual of Drums* begins with a calm and gentle introduction that culminates in a fast-paced section that intermingles with solemn chanting. Jerome Kugan (*Kakiseni*, <http://www.kakiseni.com/articles/ reviews/MDE1OQ.html>) writes in his review: 'Goh's…compositions…took the show beyond Chinese drums and into proper theatre as they explored in a rather spooky avantgarde way…interesting uses of movement…and voice.' Despite its rather drawn-out introductions and repetitions, the performance was generally well received, according to Kugan.

As with Tan's initial concept for the drum performance, Goh's Hands Percussion similarly claims a Malaysian identity. In our conversation (Interview with Bernard Goh, Kuala Lumpur, 28 June 2002), Goh was emphatic in insisting that 'the 24-drum team is a form of the Malaysian identity'. He continued by saying that Hands Percussion 'is locally formed and not from China, although we use the drums that we bought from China…Our drumsticks are not normal drumsticks that Chinese drummers use. We use *lamin* wood [hardwood] from Malaysia…The style of drumming is a local creation, not from China.'

It is interesting to note that despite statements such as 'I'm Chinese educated. I learnt the Chinese drum…I [was] born here [in Malaysia]. My father [was] born here', there seems to be a conscious or unconscious need for Goh to distinguish himself and his troupe from Mainland China. He said:

> When I perform elsewhere, like in Taiwan or in Singapore, people never think that our performance is from Malaysia. They always think that it is from China…If I perform in Australia, let's say, when they hear us, they'll know this is Chinese but it is not from China…[But] how do we identify it as coming from Malaysia? (Interview with Bernard Goh, Kuala Lumpur, 28 June 2002)

There appears to be a need to set himself and his troupe apart from other performances that are based on Chinese traditions and to 'locate' this form of performance in the 'place' of Malaysia.

> I want people to know Malaysian Chinese can do something like this, influenced by all the different cultures here. It must be very different from other Chinese who play Chinese drums. Of course, we use Chinese drums, but the way we play them is different because I was brought up here [in Malaysia]. Every night I have to go to the *mamak* store.[10] That's already part of my culture. I can speak Malay. I don't take porridge for breakfast, instead I eat *nasi lemak*.[11] I don't think we should so clearly define ourselves as 'Chinese' nowadays. We are so 'mixed', so rich in culture. (Interview with Bernard Goh, Kuala Lumpur, 28 June 2002)

Hands Percussion's web site (<http://www.hands.com.my>) also informs us that as part of their goal, the group aims to study the 'different percussion cultures to produce multi-ethnic sounds and beats in its performances', with the objective to *perform* a multicultural identity. Goh was quick to emphasise the use of local materials and the influence of other cultural traditions: 'I try to create new things, like the ways [of] beating the traditional drums, movements that my *sifu* [teacher] taught me. I learnt the *rebana*[12] and the *tabla*.[13] When they play together, and oh, that is harmonised, that's truly "Malaysian"' (Personal communication 2002).

Nonetheless, it is not so much the multicoloured range of diversity that Goh emphasises and envisages; rather, it is the hybridity that arises from the generations of intermingling of the different peoples and cultures. For Goh disputes the mere display of diversity in performance when he says 'I think we must do something beyond' playing together in harmony with the *rebana* and the *tabla*. Goh and Ch'ng's vision aims for a more nuanced understanding of the convergence of ethnicities and cultures. They conceive of a Malaysian performance as an intercultural performance—that is, to explicate the intersections of different cultures by exploring the space in between cultures. I shall return to this later in the chapter.

Between noise and music

At this juncture, I shall return to the initial aim raised at the beginning of this chapter: to chart the abstraction of the drum from its traditional context and function to its subsequent appearance on stage, and how this transplantation from a street performance to a staged performance has impacted on the *shigu* in its representation of Chinese culture in Malaysia. Before I proceed I will attempt to give an account of the full dress rehearsal that I attended. The rehearsal was held at the performance venue in a suburb of Kuala Lumpur. For this performance, there were five red drums placed in an inverted 'v' on the left side of the stage, while five black drums were placed in a circle facing each other on the right side of the stage; another three were positioned behind these. The drums at the front were positioned lower so that the drummers could sit and hit them, while the ones at the back were placed higher so that the drummers had to stand to hit them. The piece began with the drummers at the front playing in unison with occasional shouts of 'hey' and 'ho' from the players. After a short pause, a solo drummer at the front gave a signal and the drummers at the back jumped up and began to play in unison with those at the front. Here was a distinct change of strokes and rhythms, with the two groups in front playing in alternating patterns. The piece cycled through various stroke styles, striking points and rhythmic patterns, producing a variety of sound qualities. Halfway through the piece, the drummers on the front left of the stage changed position: standing with feet apart facing the audience and striking rather aggressively at the drums at an angle while moving in time to the rhythm. The players at the back moved away from the drums and lowered themselves to the ground to change instruments. They started to play handheld chimes (usually associated with Buddhist chants). Shortly after, the drummers on the front right of stage stood up and struck the drums in a similar position as the other group. The piece picked up tempo and volume; there were more shouts and vocalisations from the drummers as the piece rose into a fury. After several minutes of this, the drumming came to an abrupt end at the height of activity and volume (see Video 6.1).[14]

In the process of transplanting the *shigu* into its new art context, it has been *abstracted*: first, from a larger mixed ensemble of gongs and cymbals to an ensemble of drums playing in concert; and second, from its function as an instrument of exorcism and folk ritual to one of artistic expression. It is removed from its context as part of the lion-dance performance and its socio-religious function. As with all abstracts, however, they bear a resemblance to what they have been abstracted from. The drum ensemble therefore does serve to remind, to invoke memory, of the street lion-dance performance and by this it is not devoid of its connection to its previous context, function and role of representation.

It has also been *displaced* from its previous place of performance—the streets—and *re-placed* on the stage in a proscenium theatre. In the course of this transplantation, the *shigu* has also been *transformed* as a noise-making instrument into one of music making; the social codes associated with the street performance have been broken and, as a stage performance, new codes are imposed on it; and its role as the sonic representation of Chinese identity is also thereby called into question. As 'noise', the sound produced by the drum and its ensemble on the streets was not restrained. As noise, it had power to silence, for anyone within the listening realm either had to speak above the din or be silenced. As noise, it created a community and articulated a space (Attali 1983:6–9) within hearing distance. In its abstraction as an art form, however, and re-placement in the concert hall, the sound produced by the ensemble of drums underwent a process of organisation and moulding by 'human hands' (Hegel 1975). As Attali (1983:6) writes, 'with noise is born disorder'; however, as organised sound—that is, music—'is born power and its opposite: subversion'. Once noise undergoes a process of organisation, it takes on a different form—music—and is, thus, subverted and becomes 'the source of purpose and power' (Attali 1983:6). In this case, noise produced by the *shigu* is subverted for the purpose of communication and self-expression (*Malay Mail*, 5 April 2002), particularly for Goh and Ch'ng. It is their voice and vision exteriorised in the performances and this is clear in their communication with me, with reporters and on their web site. Their subversion of the instrument involves changes to the way the drums are placed, positioned and struck.

The ensemble's capacity to produce a high volume of sound does not suit the confined space of the concert hall, particularly when more than one drum is used. To compensate for this, Goh told me that he had to derive a number of different ways of playing the drums: the different strokes and strength levels and striking at different parts of the drums to produce softer sounds and other sound qualities—for example, striking the side of the membrane cover produces a softer and different quality from striking the drum in the centre of the cover. Striking the side of the drum, the wooden part of the body, also provides a different quality. Thus, any one piece comprises a variety of strokes, striking strengths and styles. He also said that due to the confined space of the hall, it was impossible to strike the drums with too much force, as this would have produced too much volume. Thus, players are required to restrict their stroke strengths.

In its previous incarnation, the drum was placed on its bottom while it was struck on the top membrane. While there are varying numbers of strokes and striking spots on the drum, its position is always the same. In his artistic incarnation, however, Goh has positioned the drums tilted to one side and in different placements on stage (or off, in some performances). This, according to Goh, produces a different sound quality by virtue of the fact that the sound produced is allowed to resonate. This is in contrast with its traditional placement, where the sound produced is immediately absorbed by the ground it is placed on (see Figure 6.1 and Video 6.1). This revised position requires the performers to strike the drums at an angle. To do so, the performers have to position themselves lower than in the traditional way of performance and with a slightly different stance than in the traditional manner. In addition to this, Goh has also ensured that the stance is, first, balanced, and second, aesthetically pleasing by incorporating martial arts stances and choreography. The different placements of the drums on stage (as demonstrated in the examples here) are purposeful for various reasons. The different positions and placements will produce a variety of sound qualities and sound displacements, from the louder and better resonance of those placed higher and on their side, to those tilted slightly and placed on the stage floor. By placing the drums in clusters facing towards each other, it also allows the production of different sets of sound possibilities that will resonate among themselves.

Attali (1983) also writes that sound articulates space and community. While in its previous performance context, the sound produced by the *shigu* on the streets was allowed to resonate uninhibited and undefined, in its new context, there was a necessity to contain it within the confines of the concert hall. The space and community articulated by the drum ensemble are necessarily exclusive, localised and limited to the walls of the concert hall. The community articulated is limited to the performers, stagehands and organisers and the paying audience. Only these few are privileged to witness the spectacle and experience the sound emanating from the stage. Anyone outside of those walls fortunate enough to be able to listen in is really eavesdropping. It is within these walls, then, that Goh and Ch'ng hold power, dictating to what and when the audience should listen and see. It is their directorship that commands attention from the lighting booth, the stagehands, the performers on stage and any other person involved. This power relation, however, is certainly not one sided; for on the flip side, the audience holds pre-eminence as to what it will pay to attend. As Goh says, Hands Percussion does not have any source of income other than the paying audience; it is not supported by government funding or any other funding bodies (Personal communication 2002).[15]

While traditional performance was conducted for a spirit world, to scare away 'evil spirits' and to usher in good ones, this new performance is now performed for a paying human audience. While the previous audience is unseen yet omnipresent, the new audience is seen (and heard, in terms of applause and the usual sneezes and coughs) and is held at bay to one side of the performance space (recalling that the normal performance space is the proscenium theatre). In short,

Hands Percussion's performance has transformed the drums from a ritualised context into one for entertainment, and with an exchange value. The better the show, the bigger the crowd it will draw. Where once the value was in the function the performance fulfilled, having been relinquished of it, a different value is placed on the *shigu*.

As an entertainment, the *shigu* performance involves a sonic and visual display, which explains the need for illustrative choreography. From being a sidekick in its previous performance, the *shigu* is now the main attraction. The multiple groupings of drums positioned in a variety of ways on stage provide a visual spectacle, particularly when coupled with choreographic movements. The movements employed in the performance are derived from martial arts and interpretative dance. The entire performance is quite demanding physically, as the performers are required to strike the drums as well as move in choreographed sequences across the stage. Some of their stances also seem quite awkward, requiring them to stand facing away from the drums while twisting their bodies around to strike the drums. This stance was meant to show off the costumes Goh had designed, for the costumes were significant. As Figure 6.1 shows, the costumes the performers wear do not necessarily place them culturally—in other words, the costumes are not derived from Chinese traditional costumes, as is the case with Tan's performers, who are 'clad in traditional Chinese costume with red bands on their heads and waists' (*Star*, 8 January 2002). Hands Percussion's performers are clad in sleeveless tops tied with a sash and pants in rather neutral colours—beige and khaki—which connote an association with the earth. The fabric for the top reminds one of natural fibre from coconut trees. The costume—the sleeveless top and the slits in the pants—is designed also to show off the physique of the performers, particularly when they move around on stage. In previous performances, the lion dancers' bodies were always obscured by the papier-mâché lion and the presence of the musicians (and their bodies) was masked by the thunderous sound vibrating from the instruments they played. The bodies of the musicians in comparison with the lion dancers' are less conspicuous and therefore do not have the same immediate visual attraction. Furthermore, because these musicians' performances are not based on virtuosic display, it is much less a spectacle. In contrast with this more traditional setting, the bodies of the performers in the present performance become the focus of the audience's attention. The physicality of their bodies in movement and in each execution of every stroke becomes the visual spectacle for the consumption of a paying audience.

The transplantation of the *shigu* into a different genre and performance space has inscribed on it a different value of a different system. As Deleuze and Guattari (1983) write, in the process of de-territorialisation and re-territorialisation, codes are translated and destroyed at the same time. The codes associated with the ritual the lion dance performs are carried over into the new 'territory' to some degree. These remnants of the previous incarnation are assigned an 'archaic, folkloric, or residual function' (Deleuze and Guattari 1983:245)—that is, the *shigu*'s previous ability to communicate as part of a ritual exercise is now relegated to a 'residual function', labelled as 'archaic' and 'folkloric'. It is, however, important to note also that this 'archaism' is not without value in the new 'territory'. As a source of entertainment for a paying audience, the *shigu* has entered into a different system, a capitalist system (Deleuze and Guattari 1983), which immediately identifies the instrument (and its performance) as a commodity, having exchange values.[16] The previous codes are rendered meaningless in this new system. It no longer communicates exorcism and the celebratory ushering in of blessings and prosperity; instead, it has been subverted to communicate the composer's (or in some cases, Goh and/or Ch'ng's) vision and voice and to appease a paying audience (and possibly whoever is involved in organising the performance). This is not to say that its previous 'archaism' and 'folkloric function' are of no value; on the contrary, they sell: these aspects that pass from the 'old' to the 'new' hold exchange value in an (art) economy of difference, 'exchanged as a currency' (Frith 2000:306), as evinced in the global popularity of 'world music'.

For identity

Here, I return to my earlier discussion of the role of the lion dance (and thus, by extension, the *shigu*) as a representation of Chinese culture and presence in Malaysia. As discussed earlier, the meaningful symbolism of the lion dance has changed over the years. Before the 1990s, the lion dance was the symbol of Chinese striving for recognition in Malaysia, while, since the 1990s, it has been deployed as a symbol of the Chinese community in a multicultural construction of Malaysia's political landscape. It has taken on a symbolic role of Malaysia's growing acceptance and tolerance of diversity within its political boundaries.

While the lion-dance performance is harnessed to the visual representation of Chinese-ness, the drums are its sonic representation. In its abstraction from the traditional ensemble and function and into a new ensemble of its own, the *shigu*'s representation of Chinese-ness has also changed in the process. In its previous performance context as part of a larger ensemble of performers (including the lion dancers), it was an ostensible, in-your-face display of Chinese-ness. The performance itself demanded attention, visually and sonically, and was therefore an easy and immediate symbol of Chinese identity. In this abstraction, however, the drums on their own, stripped of the lion dancers, are a toned-down version of the ostensible display of Chinese-ness. Re-contextualised in an ensemble of drums, the *shigu*, as a Chinese representation, lacks the immediacy of a visual and sonic street display of the lion-dance performance. Aside from the ethnicity of the performers and the recognisably Chinese origin of the drums, the Hands Percussion performance could easily pass for a *taiko* performance.[17]

In its new ensemble and in a concert performance space, reinscribed onto the instrument is a fractured Chinese identity. This is reflected in the mission statement on Hands Percussion's web site as well as by Goh himself: to preserve the tradition on one hand and, on the other, to transform the tradition into a modernised art form. As an art form, it is un-moored from its anchorage in ethnic difference, and this difference is flattened out and repackaged (stylised) for a universal audience. While the traditional performance of the lion dance is an ostensible display of cultural difference, which immediately projects the presence of Chinese-ness, the *shigu* ensemble as an art form does not have the same effect. Just as with the costumes and the choreography, the drum ensemble on its own bears only superficial connections to its Chinese origin: the drums themselves are recognisably those used in the lion-dance performance, the ethnicity of the performers themselves and the themes on which the performance pieces are based. The performance on the whole bears an imprint of hybridity; on the one hand, it looks and sounds 'Chinese', but on the other, it does not. In some ways, it is a 'modern' performance—that is, not traditional and stylised, yet 'archaic' at the same time. By extension, the 'Chinese' identity represented by Hands Percussion is set apart by its difference and particularity, but also bears a notion of universality and sameness. As an instrument of representation, the *shigu* serves as a palimpsest, where the 'modern' and universal are inscribed on the 'traditional' and particular.

Despite this trace of tradition in this otherwise modernised performance, Goh's mission (particularly when he talks of how 'Malaysian' these Chinese drums are) not only speaks of the Chinese contribution to the national consciousness, it is a conscious effort on his part to emphasise the Malaysian factor in this performance of seemingly Chinese origin. This is, of course, a counteraction to a very early denouncement of non-Malay contribution to the Malaysian national culture as was evinced in such claims that the lion-dance performance should not be included as part of the national culture, being deemed of 'foreign' origin. So, in this present abstraction and re-contextualisation of the *shigu* performance, Goh (and Tan as well) is consciously claiming the performance as different from its traditional role as lion-dance accompaniment. While the lion-dance performance originated from China, the *shigu* performance itself did not. Goh says, 'I just hope I can create something that belongs to us [Malaysians]. The style of drumming is a local creation, not from China…I want people to know Malaysian Chinese can do something like this, influenced by all the different cultures here' in Malaysia (Personal communication 2002).

In an interview with Jad Mahidin of *The Malay Mail* (5 April 2002), Goh said: 'Though we use mainly Chinese drums, we're not playing them the traditional way…The music will sound more Malaysian than Chinese.' Apart from the geophysical space in which Goh and the rest of the Hands Percussion team reside, the place where the concept is founded and the origins of the materials of the various parts of the ensemble and performance (drumsticks and costumes), how these Chinese drums are a 'Malaysian sound' is yet to be determined.

Perhaps 'being Malaysian' is performative, in that identity is a *productive* process. Based on J. L. Austin's (1975/2004) speech act theory, Judith Butler (1990/1999:173) argues that identity is an 'enacted fantasy' in which 'we perform *to create* who we are' as opposed to 'we are, therefore, we perform'. Thus, for Goh to claim that the *shigu* performance is 'Malaysian' does not mean that it is essentially so, but rather that the ensemble enacts their national identity with each performance. One way of *performing* Malaysian identity could be in contributing to the nation-building project. Goh said in our conversation, 'I formed a group in Taiwan. They were amazed by what we are doing. They wonder how people in Malaysia can come up with something as powerful as this' (Personal communication 2002).

Underlying this statement is very likely the consciousness of the aims of Mahathir's Vision 2020 and its slogan, '*Malaysia Boleh*' (Malaysia can). While the slogan was composed to foster a sense of confidence in the capability and capacity of Malaysia's mission of achieving developed-nation status by 2020, it has been taken up at the level of everyday living. The slogan is chanted in sports arenas and displayed on banners and its realisation has been symbolised in the construction of the Petronas Twin Towers, the breaking of Guinness World Records, and so on (see Bunnell 2004; Goh 2002; Tan 2003). Reading between the lines of what Goh said in our conversation, it is possible to conclude that he envisions Hands Percussion as participating in the nation's drive for global modernity, and thus, it *performs* its Malaysian-ness.

While the lion-dance performance has been harnessed for the display of multicultural Malaysia, as invoked in the Ministry of Culture, Arts and Tourism's *Citrawarna* parade mentioned earlier, the *shigu* performance as advocated by Hands Percussion is an appropriate conduit for the performance of the imagined community of *Bangsa Malaysia*. The lion-dance performance used in that parade and used in everyday Malaysian living renders a picture of cultural and ethnic harmony, of tolerance and acceptance, particularly in the cross-ethnic use of the lion dance, as in the case of Indian Muslim Salim Khan Kabor's annual invitation for lion-dance performances at his restaurant. Such an event has been reported positively in the local newspaper as demonstrative of a 'multicultural society' (*New Straits Times*, 1 February 2006). The *shigu* performances by Tan Fui Choong and Bernard Goh's Hands Percussion similarly promote a multicultural perspective, particularly when playing in concert with instruments such as the *gamelan* and *tabla*. Of note is not so much what is being played, but what cultural background each of the instruments stems from and therefore what ethnicity they signify. Thus, each time these various instruments are used in a concert, it is a performance (or enactment) of identity, an imagined *Bangsa Malaysia*, above and beyond the literal coming together of these instruments. It is a *performative* enactment of an identity that is yet to be realised.

Goh's vision, however, goes beyond a display of diversity. As I mentioned earlier, his conception of 'Malaysian' involves inherent hybridity as a result of the generations of intermixing of various ethnic groups living together. I have argued elsewhere (Chan 2005:33) that performance pieces such as those by Hands Percussion are ready-made spaces for the 'creative production and fabrication…of new identities'. This conception of Malaysian identity is echoed in other art forms and by fellow practitioners in Kuala Lumpur—for example, Mew Chang Tsing and her performance of *Lady White Snake* and *Re: Lady White Snake*. Mew Chang Tsing is the artistic director of RiverGrass Dance Theatre, established in 1996 and committed to 'establish[ing] a Malaysian identity that future generations can identify with' (<http://www.rivergrass.com.my/new>). Mew says she seeks to 'do something Malaysian' (<http://www.kakiseni.com/articles/reviews/MDE1OQ.html>). She defines Malaysian culture not as Malay culture but as

> a distinct blend of often separately identified cultures such as Malay, Chinese and Indian. She thinks Malaysians can be too compartmentalised when looking at themselves. Mew herself is a product of an 'inter-cultured' society—her word…She thinks Malaysians are 'inter-cultured'. 'You only have to look at our every day language and food, such as *roti canai*[18] and *laksa* to see that we are much more mixed culturally than we sometimes think.'
> (<http://www.kakiseni.com/articles/reviews/MDE1OQ.html>)

Goh shares this notion of a hybrid identity and performance, as demonstrated in the piece *Dialogue in Skin*. The piece is in five parts: 'Drumbeat Inferno', 'The Time Jungle', 'Fluency', 'Armour and Skin' and 'Centre of Gravity'. In its performance at the 2004 Penang-YTL Arts Festival, the *shigu* was joined by a Beijing opera drum, several *gendang*,[19] a *gamelan* set and *sitar*. According to a report in the local newspaper, the *New Sunday Times*, and the Hands Percussion web site, this piece attempts to investigate the notion of inter-culturalism on the basis of drumbeats, as metaphors of the 'heartbeat', 'the essence of life' (*New Sunday Times*, 6 June 2004). Here cross-culturalism is constructed on the concept of dialogue and interaction between the instruments based on a very fundamental element of humanity, the heartbeat. The performance enacts the intersections of the multiracial society. As the unnamed reporter of the *New Sunday Times* article wrote, 'it was about an active conversation, a living relationship that ignites between the skin of the drum and the touch of the drummer' (*New Sunday Times*, 6 June 2004).

Without further research and close reading of each of the performance pieces, it is impossible to ascertain in more detail what inter-culturalism means to these practitioners, Goh included. Suffice to say, they envisage an inter-cultural

Malaysian identity as more than a mere acceptance of diversity where each community is allowed its differences without external intervention. Their vision of a Malaysian identity is a hybrid mix of the various cultures in Malaysia, founded on the geophysical 'place' of the nation, accepting that it is a 'meeting place' of these various cultures.

Conclusion

This chapter set out to investigate the abstraction of the *shigu* from its traditional function as noise maker in its role as accompaniment of the lion-dance performance on the streets and its 'translation' to a modernised traditional context as art in a drum ensemble on stage. It also sought to ask if this process of 'translation' had changed its representation of the Chinese in Malaysia and, if so, how. What I have attempted to demonstrate in these past few pages is that in the process of 'translation' or transplantation, the *shigu* as an abstraction of the lion-dance performance has been subverted for artistic expression and communication by (usually) a single person—Bernard Goh—for a paying audience, and thereby, commodifying the drum and its performance. It has become a spectacle that has an exchange value placed on it. Its representation of the Chinese in Malaysia, too, has changed in the process. From a representation of Chinese culture and presence in a pro-Malay national environment, the drum ensemble has been seized to represent intercultural Malaysian identity, *Bangsa Malaysia*, albeit with a residual element of Chinese-ness. It would be interesting to note how recent events—that is, specifically the emergence of a stronger and more vocal opposition in the Malaysian Parliament—will have trickle-down effects on the lived culture and, thereby, its cultural products.

It is apparent that such an abstraction, although simple to do, is definitely not so simple in its implications. What this chapter seeks to reveal are some of the complexities and the multiple layers of meanings and significance in the simple act of removing the drums from their original setting to an artificial and artistic one. Not all, however, is changed. The sound of the *shigu* remains unmistakable and is still capable of evoking a memory of the lion dance, despite the staged performance, varying drum stokes and the new ensemble.

Bibliography

Attali, Jacques 1983, *Noise: The political economy of music*, University of Minnesota Press, Minneapolis.

Austin, J. L. 1975/2004, 'How to do things with words: lecture II', in Henry Bial (ed.), *The Performance Studies Reader*, Routledge, London and New York.

Bunnell, Tim 2004, *Malaysia, Modernity and the Multimedia Super Corridor: A critical geography of intelligent landscapes*, Routledge Curzon, London and New York.

Butler, Judith 1988, 'Performative acts and gender constitution: an essay in phenomenology and feminist theory', *Theatre Journal*, vol. 49, pp. 519–31.

Butler, Judith 1988/2004, 'Performative acts and gender constitution: an essay in phenomology and feminist theory', in Henry Bial (ed.), *The Performance Studies Reader*, Routledge, London.

Butler, Judith 1990/1999, *Gender Trouble: Feminism and the subversion of identity*, Routledge, London.

Carstens, Sharon 2005, *Histories, Cultures, Identities: Studies in Malaysian Chinese worlds*, Singapore University Press, Singapore.

Chan, Amy W. Y. 2005, Composing race and nation: intercultural music and postcolonial identities in Malaysia and Singapore, PhD dissertation, Faculty of Arts, The Australian National University, Canberra.

Crouch, Harold 1996, *Government and Society in Malaysia*, Cornell University Press, Ithaca, NY.

Deleuze, Gilles and Felix Guattari 1983, *Anti-Oedipus: Capitalism and schizophrenia*, University of Minnesota Press, Minneapolis.

Department of Statistics 2000, *Yearbook of Statistics Malaysia 2000*, Department of Statistics, Kuala Lumpur, Malaysia.

Frith, Simon 2000, 'The discourse of world music', in Georgina Born and David Hesmondhalgh (eds), *Western Music and its Others: Difference, representation, and appropriation in music*, University of California, Berkeley.

Goh, Beng-Lan 2002, *Modern Dreams: An inquiry into power, cultural production, and the cityscape in contemporary urban Penang, Malaysia*, Cornell Southeast Asia Program, New York.

Harnish, David 1998, 'Nusa Tenggara Barat', in Terry E. Miller and Sean Williams (eds), *The Garland Encyclopedia of World Music. Volume 4. Southeast Asia*, Garland Publishing, New York, pp. 762–85.

Hegel, G. W. F. 1975, *Aesthetics: Lectures on fine art*, Clarendon Press, Oxford.

Johnson, Henry 2005a, 'Performing identity, past and present: Chinese cultural performance, New Year celebrations, and the heritage industry', in Paul Millar, Keren Smith and Charles Ferrall (eds), *East by South: China in the Australasian imagination*, Victoria University Press, Wellington, pp. 217–42.

Johnson, Henry 2005b, 'Dancing with lions: (per)forming Chinese cultural identity at a New Zealand secondary school', *New Zealand Journal of Asian Studies*, vol. 7, pp. 171–86.

Kua, Kia Soong 1990, *Malaysian Cultural Policy and Democracy*, Resource and Research Centre, Kuala Lumpur.

Layton, Robert 1992, 'Traditional and contemporary art of Aboriginal Australia: two case studies', in Jeremy Coote and Anthony Shelton (eds), *Anthropology, Art and Aesthetics*, Clarendon Press, Oxford.

Loh, Francis Kok Wah 2002, 'Developmentalism and the limits of democratic discourse', in Francis Loh Kok Wah and Khoo Boo Teik (eds), *Democracy in Malaysia: Discourses and practices*, Curzon Press, Richmond, Surrey, pp. 19–50.

Marcuse, Herbert 1978, *The Aesthetic Dimension: Toward a critique of Marxist aesthetics*, Macmillan Press, London.

Nasuruddin, Mohamad Ghouse 1992, *The Malay Traditional Music*, Dewan Bahasa dan Pustaka, Kuala Lumpur.

Sani, Rustam A. 1992, *Melayu Baru: Beberapa persoalan sosio-budaya*, Institut Kauian Strategik dan Antarabangsa (ISIS) Malaysia, Kuala Lumpur.

Sarwar-Yousof, Ghulam 1986, *Ceremonial and Decorative Crafts of Penang*, The Phoenix Press, Penang, Malaysia.

Slovenz, Madeline Anita 1987, '"The year is a wild animal" lion dancing in Chinatown', *The Drama Review: TDR*, vol. 31, pp. 74–102.

Tan, Chee Beng 2001, 'Chinese in Southeast Asia and identities in a changing global context', in M. Jocelyn Armstrong, R. Warwick Armstrong and Kent Mulliner (eds), *Chinese Populations in Contemporary Southeast Asian Societies: Identities, interdependence and international influence*, Curzon, Richmond, Surrey.

Tan, Sooi Beng 1989/1990, 'The performing arts in Malaysia: state and society', *Asian Music*, vol. 21, pp. 137–71.

Tan, Sooi Beng 1992, 'Counterpoints in the performing arts', in Francis Kok Wah Loh and Joel Kahn (eds), *Fragmented Vision: Culture and politics in contemporary Malaysia*, University of Hawai'i Press, Honolulu.

Tan, Sooi Beng 1993, *Bangsawan: A social and stylistic history of popular Malay opera*, Oxford University Press, Singapore.

Tan, Sooi Beng 1994, 'From syncretism to the development of parallel cultures: Chinese–Malay interaction in Malaysia', in Stephen Blum and Margaret J. Kartomi (eds), *Music Cultures in Contact: Convergences and collisions*, Currency Press, Sydney.

Tan, Sooi Beng 2003, 'Multiculturalism or one national culture: cultural centralization and the recreation of the traditional performing arts in Malaysia', *Journal of Chinese Ritual, Theatre and Folklore*, vol. 141, pp. 237–58.

Endnotes

[1] For further information, see <http://www.genting.com.my/en/live_ent/2000/liondance/video2000.htm> (viewed 14 August 2006). A short chapter such as this cannot do justice to an investigation of the significance of the emergence of these competitions and the changes wrought on the cultural practice itself.

[2] Bumiputra can be translated as 'sons of the earth', and refers to those who are recognised as indigenous to the place—that is, including Malay and aboriginal communities.

[3] The National Culture Policy was therefore based on the following principles: the national culture of Malaysia must be based on the cultures of the people indigenous to the region; elements from other cultures that are suitable and reasonable may be incorporated into the national culture; and Islam will be an important element in the national culture (quoted in Tan:283).

[4] The 1970s and 1980s were highly wrought with politicised assertions of Malay-ness and non-Malay-ness, and this altercation took place within the representations of culture (see Carstens 2005; Kua 1990; Tan 1989/1990, 1992, 1993, 1994, 2003).

[5] Bahasa Malaysia is the Malay language adopted and adapted as the national language.

[6] The *kompang* is a single-sided membranophone that comes in several sizes and shapes (Harnish 1998:768); it is also known as the *rebana*, popular in certain states of peninsular Malaysia.

[7] It was telecast by NTV7, a local broadcasting company with coverage restricted to only the Klang Valley, which includes Wilayah Persekutuan (Federal Territory) and parts of Petaling Jaya.

[8] I was watching a delayed broadcast of the performance the next day, during working hours.

[9] For more information, see <http://www.hko.gov.hk/gts/time/24solarterms.htm> (viewed 17 August 2006).

[10] A *Mamak* store is a local feature of the Kuala Lumpur urban night scene. These 'stores' are temporary cafes set up from the early evening to near dawn the next day by Indian Muslims, specialising in *roti canai* (Indian bread) and *teh tarik* (Malaysian-style milk tea). They are patronised by locals of various ethnicities.

[11] *Nasi lemak* is another local feature. It is rice cooked in coconut milk and taken with fried anchovies and chilli and prawn paste.

[12] See note 6.

[13] The *tabla* is a set of drums usually associated with Indian music.

[14] Due to copyright issues, I am only able to show this version of the performance. Hence, Video 6.1 will not fit exactly the description earlier but it is the closest of all performance videos I have.

[15] The National Cultural Policy of 1971 dictated the funding guidelines of the incumbent ministry in charge of the arts; however, there were some exceptions to this restriction (see Chan 2005 for more information).

[16] This is not to say that the exchange value is based solely on the instrument itself; the performance being a product of labour adds value to the commodity.

[17] For information regarding *taiko* performance, see <www.eitetsu.net/>; <www.taikoz.com>; <www.taiko.org>; <www.taikoarts.com>

[18] *Roti canai* and *laksa* are popular local dishes; the former is a staple menu of the previously mentioned *mamak* store, and the latter is a spicy noodle dish.

[19] The *gendang* is a double-sided barrel-shaped membranophone in varying sizes. It is 'played with both hands, one striking each face' (Nasuruddin 1992:15).

7. Domesticating the foreign: singing salvation through translation in the Australian Catholic Chinese community

Nicholas Ng

Audio samples and video footage relating to this chapter are available online at:
http://epress.anu.edu.au/titles/sounds_translation_citation.html

Introduction: singing of a different kind

St Joseph's Chapel, Asiana Centre (Ashfield)

4 December 2004, 7.18pm

On the whole, this community is very well organised in what I see as a lively, post-Vatican II atmosphere…Hymns are sung in Cantonese and inculturation seems to have taken place at a moderate level: the Mandarin script is clearly visible in signs and religious slogans around the place…During the opening of the new Annex last year, there was a roast suckling pig that seemed to be an additional sacrifice to the usual one at Mass.

An ancestral plaque with a joss stick urn stands to one side of the old chapel. On All Souls' Day[1] and *Q ng Míng Jie*,[2] incense and joss sticks are lit with an offering of flowers and fruit. The ancestors are invoked…there is a Chinese saying: 'no ancestors, no identity'.[3]

And today, I attended the blessing of a statue of Our Lady of the Immaculate Conception…High on a Roman-style pillar, she stood pure and perfectly carved…[as] Father Tung sang the *Tota Pulchra Est* ('Thou Art Wholly Beautiful') in Latin in her honour.

Afterwards, in the community hall, the younger…generation murmured with calm reflect[ing] on what's cool and what's not in Sydney-slang. (From field notes, 2004)[4]

On 4 December 2004, hundreds of Chinese Catholics congregated for mass in the burgeoning Sydney suburb of Ashfield. As voices rose in fervent song and praise with the incense and intoned prayers, it soon became apparent that this was no regular mass. Certainly, it was a feast on a grand scale that began with the procession of white-clad altar servers following a fraternity of friars, brothers and three concelebrant priests, one of whom was non-Chinese. Most probably for his convenience and that of the smattering of non-Chinese in the congregation, the entire liturgy was conducted in English while Cantonese and Mandarin translations of the psalms, scriptural readings, homily and community announcements were provided by laypeople speaking into a microphone at the lectern.

The music performed was of special note. At times, it was unaccompanied plainchant sung in answer and response between the main celebrant and the congregation. At other times, an electric organ accompanied 'traditional'-sounding Western hymns led by the choir. In addition were hymns more contemporary in sound accompanied by organ, amplified guitar, electric bass and drums. Cantonese and English were the languages used in song and not everyone referred to the hymnals provided; bilingual hymn texts and the proceedings of the mass in Chinese traditional script shone constantly from a data projector for those who needed them. After the final blessing, the altar servers and clergy genuflected and processed out to the lively strains of *Give Thanks*, a hymn in light rock-and-roll style written by liturgical lyricist and composer Henry Smith. The choir and congregation sang this single verse and refrain song of worship through in English once and repeated it in Cantonese. Energy was high and the eruption of conversation when the music ended was indicative of the cheerful mood shared by those present.

The people had gathered to celebrate the fiftieth year of their community's establishment. This event also launched the 'Wounded Church in China' project[5] and commemorated the one hundred and fiftieth anniversary of the dogma of the Immaculate Conception of the Blessed Virgin Mary. In bright, celebratory colours, a picture of a highly stylised Asian Madonna and Child featured on the inside cover of a special souvenir edition of *Marian Art* published by The Catholic Weekly. As I stared with some wonderment at this icon, a number of questions sprang to mind: where, when and how did the intriguing corpus of music I had just heard come about? Why do these people continue to sing it? How do these people, known as the Australian Catholic Chinese Community (ACCC), maintain and preserve all three disparate identities (Australian, Catholic and Chinese) while asserting each one individually—or do they? Is it possible to be all three at the same time in what the media often portrays as the 'Chinese community' in general?

In this study, I attempt to address these questions through examining the contemporary musical performance practice of the ACCC. Through a unique music developed over more than 50 years, the community is able to assert on the one hand, its difference from other Catholics in Sydney, while on the other hand showing a willingness to adapt to its new surroundings. The community is also able to distinguish itself from other Chinese migrants in Sydney. What is often called the 'Chinese community' in Sydney is in fact a number of smaller migrant groups, each with similar values and common lifestyles. Music is closely linked to the cultural and social practices of these groups, bound either by religion, dialect and clanship or some form of common interest. The ACCC as a subgroup exhibits a unique culture in which the issue of translation is all important and has much to do with the current state of its music.[6] While this music can be classified overall as 'hybrid', it is by the same token like one of the many 'flavours' of Chinese culture that exists in the globalised world.

This study posits that members of the ACCC form part of a transnational diasporic zone in which there exist continual links between 'the local and the global, the here and the there, past and present' (Ang 2001:34). Sydney's Chinese Catholics form what Mark Slobin (1993:64) dubs a 'diasporic interculture, which emerges from the linkages that subcultures set up across national boundaries'. By examining the music of this subculture, which is primarily a vocal genre, we can see how conscious decisions in musical style and composition, in addition to choice of language, are integral in helping people identify and adjust culturally to their new adopted homeland. On the one hand, the need to remain distinct from the majority of Australian society by preserving cultural elements is strong. On the other is an urge to translate what is considered foreign in the community into something more localised in order to adapt to the wider non-Chinese community. My analysis touches on the intersection of music, language and culture. Using field data gathered periodically from 2000 to 2004 and certain established theories combined with a general knowledge and understanding of the topic, this chapter shows how this sub-community of Chinese Catholics in Sydney creates a sacred space unique to themselves through the medium of song.

Singing praise as one

> Music-making has been known to have the power to unite groups, strengthen cohesion among people within a community, and articulate ethnic identity. (Lau 2005:145)

Citing various scholars (see also Slobin 1994:245; Stokes 1994; Hosokawa 1998; Nettl 1983:162–5; Radano and Bohlman 2000:8; Averill 1994) in an article on Chinese amateur singing in Bangkok, Lau (2005:145) reminds us of the phenomenon in which music in the form of national anthems, songs of patriotism, military bands and national orchestras has often been used symbolically to create national and diasporic bonds. In the case of Sydney's Catholic Chinese, this musical adhesive functions to unite people in the worship of a single deity through poetic song texts describing aspects of their belief system. These texts, if not liturgically or scripturally related or inspired, can refer to some matter of religiosity and even instil a sense of nostalgia for an imagined homeland.

Hymn singing is not necessarily like the singing of national anthems, the words of which people have often 'remembered well enough to forget to understand'

(Kelen 2003:162). Far from a phatic exercise, hymns and plainchant singing in one's native tongue communicates a sense of belonging and can be as strong a symbol of unity as it is a binding force in community groups worldwide. Hymns and plainchant sung during mass not only help reaffirm what people believe in, they help express a sense of collective identity as they sing to themselves and one another about what they hold dear to their hearts. Even in the most impersonal of church situations in which community members are acquainted on a superficial level limited to weekly contact, congregational singing still has the means to help people identify as a group. According to Frith (1996:121), song, or more precisely the act of responding to song, draws audiences into emotional alliances with performers. In the case of the ACCC, in the context of liturgical singing, I propose that the singing of sacred songs helps unify the group as one, in which social status, age, gender and other differences are forgotten in the music.

In his study of karaoke singing, Hiroshi Ogawa (1993) observes that the people present are enclosed within the walls of a 'karaoke space'. With everyone considered friends in this space, 'a person singing in the presence of others in spite of shyness is thought to be trusted. Both sharing a "karaoke space" and singing in the presence of others reinforce group consciousness' (Ogawa 1993:2). Although hymns are quite a different thing to karaoke, both preoccupations involve singing, and the same fundamentally human concept of 'group consciousness' applies. It is a consciousness in which many things are held in common and are essentially communicated. While any two people of any two societies can arguably be able to find things in common and locate a deferential point of identification, people who sing during mass identify at many levels because they are of the same community and are able to communicate the many things they hold in common. 'What they must have in common…are aims, beliefs, aspirations, knowledge—a common understanding—likemindedness as sociologists say' (Dewey 1916:5–6).

Casey Man Kong Lum (1996:21) believes that 'likemindedness among people gives them the identity of being members of a community. It is this likemindedness among people that forms their common frame of reference to construct and comprehend their social and cultural reality.' For Catholic Chinese in Sydney, this reality is one of belonging to a religion as an ethnic group in a multicultural nation in which their ethnicity separates them from society as 'different'. They possess a common belief system and culture easily communicated through the music of their community. Furthermore, Christopher Small (1998:9) maintains that the making of music is a convenient channel through which an individual or groups of individuals express or cultivate their identity in relation to their social environment. For Chinese Catholics, apart from using hymns as an expression of faith, singing sacred music is a powerful and evocative way to assert a sense of Chinese-ness in community groups in the homeland and in new areas of settlement within the diaspora. Undoubtedly, hymns and plainchant sung in one's native tongue are rich in religious and ethnic meaning. Although it is not possible for me to write on a general basis for all Chinese Catholic communities worldwide, it is quite certain that the religious culture of these groups is subject to change with the onset of migration. This will be shown in the following discussion using the case study of the ACCC in Sydney.

Singing praise from the home front and beyond

The families and religious workers who established the beginnings of the ACCC in Sydney originated mostly from the southern Chinese province of Guangdong and what is now the Hong Kong Semi-Autonomous Region. Audrey Donnithorne (Personal communication, 10 May 2006, 13 May 2006) remembers that these émigrés came from a practice of sacred music that involved plainchant in Latin (synonymous with Chinese Catholic worship for many centuries) and hymn singing taken from plainchant and other genres (a more recent practice introduced by missionaries from the 1800s). Catholic hymnody today is unlike that of the Latin hymns of the past.[7] The present-day hymnody was developed to counter the Protestant hymn movement. Until the Second Vatican Council (1962–65), all sacramental celebrations including the Holy Mass were conducted in Latin, according to Father Bonaventure Tung (2006). With the reforms of Vatican II, during which 'inculturation'[8] was encouraged, the devout in Guangdong and Hong Kong started to worship and sing in Cantonese while Latin fell into disuse. In the colonial climate of Hong Kong, spoken and sung English were also introduced to worship. By the 1970s, Chinese locals had domesticated and made 'native' the monotheistic religious culture from Europe that they had embraced in the search for salvation. Their foreign Roman God was now made local and many Chinese soon found Catholicism closer to home with the liturgy and other sacred texts translated into their native tongue. Not all condoned the changes made. To the dismay of many traditionalists, classical hymn texts in archaic Chinese verse were rewritten in everyday language while tangible expressions of religious culture such as statues, holy picture cards and other ecclesiastical paraphernalia[9] were increasingly Sinicised. Throughout China, Hong Kong and Taiwan, local dialects and languages were employed to unite the people in song during worship by translating Western hymns and even Western folk songs with religious texts into the vernacular, and by the composition of contemporary hymns.

Such was the unifying socio-religious practice that was brought to Sydney from 1954 with the invitation of Sir Norman Thomas Cardinal Gilroy (then Archbishop of Sydney) for Father Pascal Chang and the late Bishop Leonard Hsu to launch the Sydney Catholic Chinese Mission.[10] In 1989, the umbrella organisation of the ACCC was officially formed as a board of trustees and as a central and organising entity for its three primary branches (Table 7.1).

Table 7.1 Australian Catholic Chinese community branch structure (1989)

Branch	Established	Location	Choir	Main language	Countries of origin
Catholic Chinese Pastoral Centre (CCPC)	1954 (moved to present location in 1985)	Crypt of St Mary's Cathedral St Peter Julian's Church, Chinatown	CCPC choir	Cantonese	Hong Kong
Asiana Centre Association (ACA)	1973	Asiana Centre, Ashfield	CCPC choir	Cantonese	Hong Kong
Western Sydney Catholic Chinese Community (WSCCC)	1990	St Dominic's Church, Flemington	WSCCC choir	Mandarin	Taiwan, Laos, Cambodia, Vietnam, Singapore, Malaysia and Timor-Leste

Source: Summarised from <http://www.chinese.sydney.catholic.org.au/E_Accc.htm> (viewed 7 May 2005).

CCPC, ACA and WSCCC

Under the pastoral care of Father Bonaventure Tung, the current Chinese migrant chaplain, the three branches of the ACCC combine to provide welfare and ecumenical services in addition to community activities for Chinese Catholics in Sydney who choose to meet, sing and worship with others of the same ethnic and religious orientation.[11] As recorded in Figure 7.1, the Catholic Chinese Pastoral Centre (CCPC) and Asiana Centre Association (ACA) in the city and eastern suburbs form a meeting ground for what is essentially a Cantonese congregation of expatriate Hong Kong Catholics and their descendents. Further west at St Dominic's, the Western Sydney Catholic Chinese Community (WSCCC) is a highly multinational community that has coalesced as a *huawen* (Mandarin-language) unit due to the plethora of mother tongues spoken at home. These Catholics meet under the spiritual directorship of Father Paul McGee from the Columban Mission Institute. In general, the culture of the most populous Chinese subgroup tends to prevail over the wider community. Hence, because Hong Kong migrants are greater in number than Mandarin speakers from Taiwan and wider Asia, Hong Kong culture appears strong in the ACCC at large. Meanwhile, Taiwanese music, food and general presentation tend to dominate the Mandarin-speaking congregation (WSCCC). Language barriers and subcultural marginalisation are, however, seen as minor issues because community members in their respective chosen subgroups are happily able to sing, worship, communicate and read in the vernacular, which in traditional script can be read in either Cantonese or Mandarin. Children are sent to Sunday school and many attend the Asiana Centre Catholic Chinese School where Cantonese or Mandarin are taught.

Figure 7.1 Sydney Chinese Catholic Mission pamphlet

澳洲 雪梨 天主教華人堂區
Sydney Catholic Chinese Community

雪梨天主教華人牧靈辦事處
Sydney Catholic Chinese Pastoral Office
432 Sussex St., Haymarket 2000 Phone: (02) 212-6823

The Sydney Catholic Chinese community is essentially a support group for migrants, which many of their children are not. Father McGee (Interviews, 4 October 2000, 15 October 2000; Phone communication, 7 April 2004), who administers the Mandarin congregation in the western suburbs of Sydney, observes that most migrant children go to school and receive a tertiary education, then leave the community when they find a job. An example of this is the Chungyue children at St Dominic's. Adelino and Iana Chungyue (Interview, 3 September 2000), a Chinese couple from Timor-Leste, say their children used to play the violin and organ at mass but now have found full-time work and are consequently 'too busy'. There are many similar cases to this happening today. The community is, however, still flourishing, with young children to university-level people currently living under their parents' influence. English speaking and bicultural, they could be there 'because Dad will kill me' otherwise, but for the present, they are building close bonds established with peers from an early age. Many, especially the children of community executive board members, are altar servers (male and female) and acolytes and participate in activities such as camps at St Francisville (the community's place of retreat) and other church-related interest and religious groups such as the St Louis Group. There have even been evangelical groups of young people who take to the streets to 'spread the Good News' to Chinese and non-Chinese with information pamphlets about the ACCC. Some even take pride in their 'Chinese-ness' and stay with the community for that very reason. 'We all grow up with *something* but we can choose *anything* by way of expressive culture' (Slobin 1993:55).

In the very way described by Slobin in the quote above, community members I have worked with certainly choose to live Catholic lives among people of the same ethnic origin. Together, they sing, eat and speak 'Chinese' in their worship of Christ. Instead of going to mainstream English mass, some community members travel long distances to Chinese Catholic centres in Sydney for religious and cultural purposes. When asked why they prefer meeting and worshipping with other Chinese, the general response from community members is that the familiarity of it all makes them 'feel comfortable'. Further comfort comes from singing and speaking in a Chinese tongue. This enables participants to identify as Chinese. Clearly, many Catholic Chinese, especially recent and elderly migrants, are affected by having to adjust to their new surroundings. The remedy to this sense of displacement is quick and easy: finding solace in the church community, which apart from offering spiritual nourishment provides them with social and mental relief from the (quite possibly externally and internally imposed) stigma of being foreign simply by being with other people who are equally foreign. In mainstream Australia, Chinese Catholics cannot help but find that they and their culture are starkly different from other Catholics and, to a certain extent, other Chinese. Catholicism existed in Sydney before their arrival, but theirs is a unique *Chinese* Catholic culture of a different hue, and of a different temperament that appears very foreign despite its highly domesticated and localised nature in the homeland. Their language, their Sinicised paraphernalia and their music are all attributes of difference that mark them as distinct from other non-Chinese Catholics in Sydney. To this day, members of the Catholic Chinese community in Sydney practice with great fervour a European religious way of life imbued with native Chinese elements, which in the old country was the result of years of adaptation into the local Chinese culture. It certainly is a vibrant, post-Vatican II atmosphere that exists today.

Figure 7.2 Suckling pig

Figure 7.3 Ancestral altar

Figure 7.4 Father Tung with Our Lady of the Immaculate Conception

'"Chineseness" is not an immutable set of beliefs and practices, but a process which captures a wide range of emotions and states of being' (Siu 1993:19). Edith Lo, who has been a choir member in the Cantonese-speaking congregation since 1985, believes it is important to exhibit her community's Chinese-ness at large events that are exposed to the public eye. Chinese New Year mass is one example. Each New Year, an ancient Chinese rite for the veneration of ancestors is practised. This ritual has been reintroduced into Chinese Catholic communities worldwide after several centuries of debate in the Vatican, which initially barred the practice (Interview with Edith Lo, 10 September 2000). Taking part in such rituals today allows the Chinese to celebrate an ethnicity that is different to the mainstream; it is something the Chinese Catholics in Sydney do with a certain pride. The picture of the Chinese Madonna and Child mentioned at the start of this chapter was vastly unlike any of the European presentations of the Virgin in the special souvenir edition of *Marian Art* published by The Catholic Weekly. Beneath the picture were the words 'Dedicated to Our Lady by the Australian Catholic Chinese Community to commemorate the celebration'.

Figure 7.5 Chinese Madonna and Child

Dedicated to Our Lady

by the Australian Catholic Chinese Community

to commemorate the celebration (December, 2004)

- The 50th Anniversary of The Franciscan Chinese Mission in Australia established by Fr Paschal Chang OFM and the late Bishop Leonard Hsu in March 1954;
- The Sesquicentenary of the Promulgation of the Dogma of the Immaculate Conception;
- The blessing of the Statue of Mary Immaculate Conception at Asiana Centre Ashfield;
- Launching of the Catholic China Mission Service Project by Australian Franciscan Provincial Very Rev Fr Stephen Bliss OFM.

The Project is to help the Wounded Church in China. The detailed plan, especially concerning recruitment of vocations, is described in a booklet endorsed by His Eminence Cardinal George Pell. The booklet is freely available to all who are interested. Please send your name and address to:

THE FRANCISCANS
38 Chandos Street
Ashfield NSW 2131 Australia
Tel: (61)2 9799 2423
Fax: (61)2 9798 6866
Email: gboggs@franciscans.org.au

In such representations from the media, there is no shame in appearing as a distinctly different group with a different history, of which 50 years on Australian soil is worth celebrating. At the community's golden anniversary, as with other large events to which important outsiders are invited, 'Chinese-ness' is exhibited in language, native customs and music through the medium of the mass. The use of English in hymns and parts of the liturgy, however, ensures that visitors from the outside are not entirely alienated from the proceedings. Hence, it is not simply a *Chinese* Catholic culture that is exhibited but a new, modified one in which we can observe the preservation of the old homeland culture and their attempts to adjust, reach out and relate to the wider Australian community as Australian Chinese. This can be seen most clearly in the sacred songs they sing today and their approach to language use.

Singing praise in the new homeland

The hymnals used by the ACCC were published in the 1980s after the wave of Sinicisation and inculturation of the 1960s and 1970s. The CCPC choir (established in 1985; see Figure 7.6) and the WSCCC choir (established c1998–99) thus lead their respective congregations in song during mass from hymnals that facilitate the preservation of pre-existing musical traditions. There are two main hymnals in current use: one from Hong Kong and another from Taiwan. World-renowned liturgist and composer Father Lucien Deiss played an immensely influential part in the development of the Catholic Chinese hymnody. Cantonese-speaking ACA and CCPC members sing from the Hong Kong-published hymnal, which exists in two volumes. This hymnal employs a combination of Western and cipher music notation[12] in the Pew Edition (see Music 7.1), while the Organ Edition is entirely in Western staff notation (see Music 7.2). Meanwhile, the WSCCC sings from a Taiwan-published Mandarin hymnal in cipher notation only, with guitar chords stipulated above the easy-to-learn melodies (see Music 7.3).[13] Following are three excerpts taken from religious services from all three branches of the wider ACCC that are very much in the 'old' tradition of music for worship. Traditionally performed in the Western world, these musical items have been translated into Mandarin and Cantonese for use during mass and other Roman Catholic rites in the community.

Music 7.1 *Huáng Huáng Shèng Ti*

171　　　　皇 皇 聖 體 (一)

```
1. 信友請來　讚美上　主，稱　頌奧妙之聖體。
2. 自作犧牲　賜與我　等，生　於聖潔童貞女。
3. 受難前夕　最後晚　餐，偕　諸兄弟齊坐席。
4. 天主聖子　一言甫　出，麵　餅即化成聖體。

5. 皇皇聖體　尊高無　比，我　們俯首致欽崇；
6. 讚美聖父　讚美聖　子，歡　欣踴躍來主前；
```

```
1. 又請稱頌至聖寶血，　主因贖世盡傾流。
2. 在世居住愛情昭著，　福音廣佈苦不辭。
3. 徹底遵守古教規禮，　巴斯卦節食羔羊。
4. 葡萄酒亦成為聖血，　此中奧妙洞悉難。

5. 古教舊禮已成陳跡，　新約禮儀繼聖功；
6. 歌頌救主凱旋勝利，　頌揚主德浩無邊；
```

```
1. 萬邦之王生於聖母，　自甘流血　為　人類。
2. 凡數十年歷盡艱辛，　立斯聖事　顯　奇跡。
3. 還要親自將己體血，　分給門徒　作　粮食。
4. 為使爾能信心堅固，　只須藉伙　活　信德。

5. 五官之力有所不及，　應由信德　來　補充。
6. 聖神發自聖父聖子，　同尊同榮　同　威嚴。亞孟。
```

（司）　主，你賜給我們天國的食粮。（衆）它給我們帶　來了喜樂。

請衆同禱：主，你在這奇妙的聖事內，留給了我們受難的紀念；
　　　　　求你幫助我們虔誠欽崇你體血的神聖奧跡，俾能常常
　　　　　領受你救贖的功效。你永生永王。　　　衆：亞孟。

1 Sing, my tongue, the — Saviour's glory, of— his Flesh the mystery sing:
   Of the Blood, all price exceeding, Shed by our immortal King,
   Destined, for the world's redemption, From a noble–womb–to spring.

2 Of a pure and–spotless Virgin, Born–for us on earth below,
   He, as man, with man conversing, Stayed, the seeds of truth to sow;
   Then he closed in solemn order, Wondrously his–life–of woe.

3 On the night of–that Last Supper, Seated with his chosen band,
   He, the Paschal victim eating, First fulfils the Law's command;
   Then as food to his apostles, Gives himself with–his–own hand.

4 Word made flesh, the–bread of nature, By–his word to flesh he turns;
   Wine into his blood he changes: What though sense no change discerns?
   Only be the heart in earnest, Faith her lesson quickly learns.

172 TANTUM ERGO

5 Come, adore this–wondrous Presence! Bow–to Christ, the source of grace!
   Here is kept the ancient promise, Of God's earthly dwelling–place!
   Sight is blind before God's glory, Faith alone may–see–his face!

6 Glory be to–God the Father! Praise–to his coequal Son!
   Adoration to the Spirit, Bond of love, in Godhead one.
   Blest be God by all creation, Joyously while–ages run! Amen.

   V. You have given your people bread from heaven. (Alleluia)
   R. The bread which is full of all goodness. (Alleluia)

Music 7.2 *Fāng Jì Gé Dě Qí Dao*

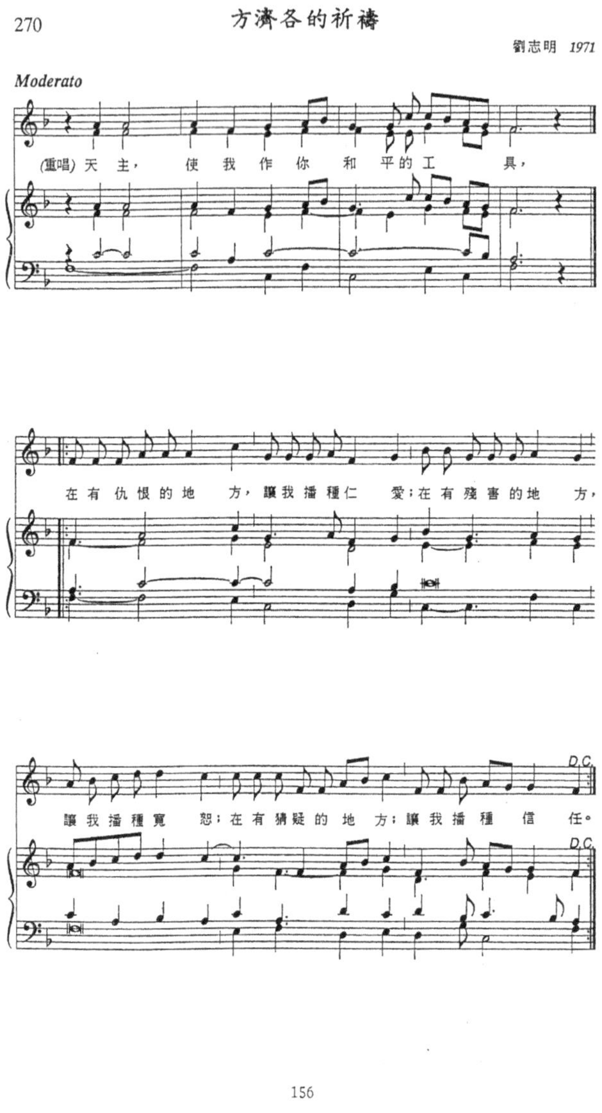

我不企求他人的安慰，只求安慰他人；

我不企求他人的愛護，只求愛護他人。

因為在施捨他人時，我們接受施予；因為在寬恕他人時，

我們獲得寬恕；因為在喪失生命時，我們生於永恆。

157

Music 7.3 *Gāo Yáng Zhàn*

Figure 7.6 The CCPC choir

The *Doxology of Praise* 'Through Him, with Him, in Him' (see Video 7.1) was intoned by Father Tung to conclude the Eucharistic Prayer during Easter Sunday mass in 2003. It was faithful to the original plainchant tune but sung in Mandarin. There are different points of reference, association and meaning here. The priest was singing in Mandarin, which he used in Cantonese and Mandarin communities. The Cantonese group might find this still slightly foreign and could only try localising this foreignness by having translators during mass. For them, God is still someone they cannot quite understand for neither of the Chinese pastors (Father Tung and Father Chang) speaks any Cantonese. At the twentieth anniversary of the Cantonese choir's establishment in July 2005, a play written by community members was enacted in which the people heard the voice of God, which was in fact the recording of Father Chang's voice in Mandarin.

The *Pange Lingua Gloriosi* is another example of plainchant (see Video 7.2). More specifically, it is a plainchant hymn, the melody of which is supported in this instance by basic diatonic chords (primarily I, IV and V). During a Saturday-night novena, the ACA and CCPC congregations sang in Cantonese in adoration of the Blessed Sacrament (see Audio 7.1). According to living memory, *Pange Lingua Gloriosi* was sung in the vernacular even before the reforms of the Second Vatican Council in the 1960s. This Latin hymn is one of the oldest in Catholicism and can be traced to the Middle Ages; its text was composed by the great St Thomas Aquinas (1225–74) for the Solemnity of Corpus Christi. It is considered one of the seven-greatest hymns of the Church (Martin 1998). In contrast, the *Doxology of Praise* would have been in Latin until the changes encouraged by Vatican II took place, as it was part of the liturgy. Plainchant, accompanied and unaccompanied, is an example of sacred music performed in the Chinese community of Sydney.

P Shì Tóng H an Chàng is a traditional hymn many in Christendom know as *Thine Be the Glory* from Handel's oratorio *Judas Maccabaeus* (see Video 7.3). In this Chinese rendition, the original melody and organ accompaniment have been retained with text in Mandarin on the glories of the risen Christ. *P Shì Tóng H an Chàng* is an example of the many hymns in Taiwanese and Cantonese hymnals that are essentially translations of hymns from the corpus of Catholic hymnody developed in the 1800s and early 1900s.

The musical items in Videos 7.1, 7.2 and 7.3 reflect what can be seen as an initial step towards the development of Chinese Catholic music with translation alone as the main focus. At this early stage, there is not much in the way of inculturation or assertion of a Chinese identity apart from language use. In contrast with the above are hymns that depart from what we might call the 'traditional' and 'conventional' in Roman Catholic congregational singing (the development of which has been influenced largely by Protestant hymns) to musical items that are more localised and innovative, a move fostered worldwide by Vatican II.

During Easter Sunday mass, the congregation of the WSCCC sings (in Mandarin) 'Lamb of God, You have taken away the sin of the world, have mercy on us' to the accompaniment of the electric organ (Video 7.4 and Audio 7.3). From my analysis of the ACA/CCPC and WSCCC hymnals, this is an interesting example of Catholic Chinese music developed in the 1970s that is still very much in vogue today. The text is set to a melody based entirely on a pentatonic scale often associated with 'Chinese music' (D, E, F#, A, B). According to community members, this gives it a 'Chinese sound'. Chinese music is generally not chorded but operates on the principal of heterophony. In this particular hymn, however, the composer has supported the melody with chordal accompaniment borrowed from Western music, the chief characteristics of which are widely acknowledged as harmony and counterpoint. According to my informants, weaving Chinese melodies over diatonic Western chords was a way of integrating a sense of native Chinese identity with the foreign musical culture of their religion. Looking through the community's hymnals, I found this to be the case for many twentieth-century Chinese hymns. Today, the efforts of inculturation in the 1970s seem to have paid off: items such as *Gāo Yáng Zhàn* are still sung in the Sydney community. There are also more hymns being composed in this fashion. Edith Lo (Interview, 10 September 2000), former president of the CCPC choir, says that hymns today are increasingly Chinese sounding with the use of 'Chinese tunes'. I am assuming that she also means hymns composed in a way similar to *Gāo Yáng Zhàn*—that is, supported by diatonic chords. In *Gāo Yáng Zhàn*, the selection of chords is the tonic, subdominant and dominant—what trained musicians might regard a simple selection of chords, no doubt, but this is for a good purpose: members of the community tell me the hymns in their community are designed for easy learning and convenience.

When writing Chinese hymns, it is important to take into consideration not so much metre but the tonal nature of the language or dialect concerned. Eddie and Annie Ho (Interview, 31 July 2005), two longstanding members of the CCPC choir, assert that writing Cantonese hymns is no simple exercise due to the dialect's nine tones as opposed to the four of standard Mandarin.[14] As a result, simply singing Western hymns translated into Cantonese, or indeed translated from any language or dialect, often leads to confusion. An embarrassing situation involves the Cantonese word '*jyu3*' ('God') (Audio 7.1) sung on the high note of '*jyu1*' ('pig') (Audio 7.2). For this reason, modern compositions by native Cantonese composers are highly valued in the old country and in Sydney as well, in the continuing interest of composing hymns in Chinese. In the homeland, it is part of the domesticating mission fuelled by Vatican II; in Sydney, it helps assert a Chinese identity and is what helps define cultural and ethnic boundaries between their community, mainstream Catholics and others.

An example of a local Sydney product using Cantonese is a composition by Eddie Ho (Music 7.4). It was written for the *Hé H o* ('Reconciliation') 2000 project from which many compositions in English and Cantonese emerged, in addition to the production of a CD, *Aì G Rén Sh ng Lù* (*Music Lovers' Life Journey*) (Figure 7.7 and Audio 7.3). Ho's composition was the winning piece in the liturgical music competition that was run that year. He is conscious that it is a 'Western' composition, which, according to him and his wife, was 'inspired by the Holy Spirit' and completed mostly in the shower over several weeks. A fan of evangelical, processional/recessional-type hymns, Ho designed it to be easily learnt for it was essentially a 'children's marching song'.

With its triadic harmony and step-wise melody, this hymn is, as Ho describes, Western in every respect, with the exception of the Chinese text, which is in contemporary (though formal) Cantonese. This hymn can also be sung in English and exists in both versions on the CD *Aì G Rén Sh ng Lù*. The single verse celebrates the intervention of Jesus in our lives of the past, present and future. It is partially related to the liturgy: the first part of the chorus is based on the *Doxology of Praise* from the Eucharistic Prayer while the remaining words call on Christians 'to brave the world and march forward under the grace of the Triune God' (Interviews with Eddie Ho, 8 December 2005, 14 December 2005).[15] Compositions such as Ho's are well regarded in the Sydney community not least because the melodic contour suits the Cantonese language perfectly without any need for altering the original text when transliterated from Mandarin, or translating it from a foreign language—a constant problem for the choir. Eddie and Annie believe that much of the beauty in singing sacred Chinese texts translated from European languages is 'lost in translation'. In Annie's opinion, the English language is not nearly as expressive as Cantonese or Mandarin, which in poetic form are elegant in their diminutive, metric nature (Interview with Annie Ho, 2005).

Eddie's piece very much complies with the Second Vatican Council's mission to indigenise the Catholic faith worldwide. Although it was composed in Sydney, this composition could be included with the preceding excerpts that exemplified the sacred music developed by the Chinese on home territory in localising their faith from a foreign belief system to one more domesticated. As mentioned earlier, this localisation was achieved through certain choices made in language, musical style and the writing of new compositions. With migration to Sydney where, to borrow Anderson's (1991) terms, 'host country' gradually turned into the 'new homeland', language and style once again became issues of consideration in the production of music. For a migrant group on alien ground, it had become evident that what was once native in the home country had reverted back to being foreign, or at least could be seen as foreign in light of the Anglo-Celtic mainstream culture.

Music 7.4 *Yéh Sū Yíng Dǎo Gùo Chì*

耶穌永常在
Jesus Forever Reigns

何潤發
Eddie Ho

耶　穌　引　領　過　去，　耶
Je - sus　led　our　yes- ter- day,　Je -

穌　掌　管　未　來，　耶　穌　也　看　顧著
sus　holds　our　to- mor- row,　Je - sus　he　cares　for our

今　天，　耶　穌　永　常　在. 耶　穌　永　常
to - day,　Je - sus　for -e - ver　reigns. Je - sus　for -e - ver

在. 　借　同　基　督，　在　聖　神　內，
reigns. 　Through　him,　with him,　in the　Spir - it,

歸　向　天　父.　在　聖　三　慈　愛　的
praise　to the Fa - ther.　Armed with love　and　grace　of the

看　顧　之　下，　基　督　徒　向　前　邁　進!
Tri - une　God,　Christ- ians　for- ward march- ing　on!

Figure 7.7 *Aì Gē Rén Shēng Lù*

Hymns such as *Yéh S Yíng D o Gùo Chì*, written for the community's *Aì G Rén Sh ng Lù* CD, can be seen as a reaction to this sense of foreignness in its superimposition of Western music (in melody and harmony) in a bilingual setting (Cantonese and English). On the one hand, it is perfectly suited to being sung in Cantonese, making it exclusively a composition for the composer's immediate Cantonese-speaking community of the ACA and CCPC. The hymn can be considered a representative item for the community, hence assisting it (like the Chinese Madonna and Child) in appearing different from the mainstream. On the other hand, the situation seems remarkably inverted when sung in English, because it can then almost be considered a local product of a community fully assimilated and adapted to its local English-speaking surroundings. Hence, Eddie's composition is able to translate group identity in setting up a barrier of difference as well as in helping the community assimilate and cater to their predominantly English-speaking children.

Chinese communities across the globe have for centuries learned to adapt, accommodate and assimilate wherever they go with the establishment of a strong economic base and a community centre (Heilbron 1998:xiv). The migrants who make up the Sydney Catholic Chinese community are no different. According to one informant, there is always in the migration process an old Chinese mode of conduct or 'etiquette' at work, the origins of which can be found in Confucianism and which comes close to the English 'when in Rome, do as the Romans do'. In the view of this informant and many others, it is important to respect and learn the ways of the locals you live among without forgetting to retain one's own language, for, as Victoria Ng believes, it is largely through language that culture is preserved (Interview with Victoria Ng, 27 May 2005). Community members in the vast ethnic arena of suburban Sydney have started to live lives of 'multiple identities' in the effort to adapt to the new culture and preserve the old with 'ongoing and regular interaction between local cultures' (Chun 1996:96).

For Chinese Catholics who live as a pocketed community in mainstream Sydney society, interaction with non-Chinese means locating ways for adapting and relating to the wider community. Bringing in English and music used in mainstream English mass was the principle means to this end. During the community's fiftieth anniversary mass in Sydney, I sang along in English to a well-known Australian mass setting by Brother Colin Smith and the 'Our Father'. According to Zita Ho (Interviews, 30 November 2004, 3 December 2004, 29 June 2005, 31 July 2005, 6 December 2005, 9 December 2005 and 12 December 2005), the use of the English language and Australian content served the specific purpose of accommodating the non-Chinese chief celebrant and visitors there. Having non-Chinese visitors is not the only incentive for the presence of the English language at mass. Many younger-generation members today, particularly those born in Australia, might be brought up 'Chinese' at home, but learn to adjust culturally to life outdoors, which in mainstream Anglo-Australian society is highly 'Western'. These children are bilingual and more often than not English is their language of choice. Parents in recent years have found themselves addressing their children in English despite spending hard-earned money sending them to Chinese-language classes. Many children, especially those who have reached adulthood, are beginning to find little relevance in the community, which operates around Chinese mass—that is, mass in a Chinese language.

For those children who do not speak any Chinese, who find little point in it or, by association, find it antiquated and not very 'cool', introducing English into the Chinese community has served a very pragmatic function. It appeals to the community's children, who are 'Aussie now', and it attracts and accommodates people from outside the community who for various reasons attend Chinese mass. Almost ironically, while their predecessors 40 years ago found joy in introducing Chinese into what was once a Latin ceremony, many now find it necessary to incorporate some English in the Chinese mass community. There has been for several years now an English children's mass and a catechism class held adjacent to the regular Mandarin services in the school hall at St Dominic's Church. There are also catechism classes in Mandarin, Cantonese and English at St Peter Julian's Church. At important events and during mass for the grown-ups and the community at large, in which there are non-Chinese celebrants presiding over the entire community, parts of the mass are sung in English and there is at least one English hymn, which is normally the *Recessional* (exit) hymn.

Such was the linguistic situation at the community's fiftieth anniversary celebration in 2004. To cater for migrant children for whom singing in English had more preferential and symbolic weight, there were additional English hymns taken from some local, but modern hymnals.

As Mass ends, *Give Thanks* (Music 7.5) is sung in English and repeated once through in Cantonese. This presents quite an interesting bilingual situation stemming from the aim of catering for non-Chinese people in the congregation and clergy by using a modern hymn in which everyone, from within and outside the community, can participate. This hymn is performed with the accompaniment of electric organ, amplified guitar, bass guitar and drums. With the influence of popular music, hymns in English signify the process of domesticating once again in the new home country by adopting the local language. *Give Thanks* as a 'pop hymn' complies fully with the Vatican's *Comme le prévoit* (CLP) of 1969, which instructs that liturgical language (more particularly the language of translation) 'should be that in common usage'—that is, language 'suited to the greater number of faithful who speak it in everyday use'.[16] I believe that this composition achieves in full what Eddie Ho's hymn accomplishes in making the music meaningful to people of Chinese and English-language backgrounds. In the case of such bilingual hymn singing, the use of English is more a need than a cultural or aesthetic choice in light of the as yet uncertain survival of native Chinese tongues in Sydney.

Music 7.5 *Give Thanks*

Singing inter-culture

The music of the ACCC presents an interesting case study of a *'hybrid* cultural form(s)' (Ang 2001:35) conceived out of a productive, creative syncretism for the consumption and enjoyment of the congregation. We can see in the community's music what Ang (2001) calls a *creative* tension between past and present—that is, between 'where you're from' and 'where you're at', or 'what we might become', as Stuart Hall (1996:4) writes.

Catholic Chinese music in Sydney appears to reflect and even emulate the phases of musical development in the Western Church. This is particularly the case in some of the older repertoire introduced to Catholic Chinese communities before migration to Australia. Today, Catholic Chinese music is still affected by changes in the modern Church since Vatican II.[17] What have directly affected and influenced the music that is still performed by Sydney's Chinese Catholics, however, are unaccompanied plainchant, accompanied plainchant and the 'traditional' hymns of the past with the accompaniment of the electric (church) organ. In recent decades, modern compositions, too, have conditioned the music at Chinese mass. The audiovisual examples above show hymns that are Western in nature and some that are more Chinese in orientation. Popular song genres such as the rock ballad have even been used as a model for hymns. Today, the electric organ and Western band instruments are used to accompany 'traditional' and 'modern' hymns, which are sung entirely in Mandarin, Cantonese, English or in a bilingual English/Cantonese alternating verse setting.

Unique to the community today is the composition of new Chinese hymns in Western style, the generational push for English and the trilingual presentation of the mass in Mandarin, Cantonese and English. This presents a case not so much of biculturality—that in-between space in which the individual is lost in the cultural translation from one side to the other of 'where you're from' and 'where you're at'—as one of hybridity, which Ang (2001:34–5) describes as that multi-perspective productivity filling the space of in-between-ness with 'new forms of culture at the collision of the two': in this case, the Australian and the Chinese, if we view it as such.

In the 'collision' of cultures, the penetration of English is inevitable and its influence continues to grow. While Chinese migrants worked hard to preserve the native tongue at home and at church in promoting their Chinese-ness as different to other local Catholics, they also sought to adapt to local customs through the English language instead of borrowing from established and internationally recognisable traits of 'Australian-ness'. In the Chinese community, English is regularly used in certain parts of the mass liturgy. It is also heard in the singing of hymns from modern Australian hymnals. There is an English version of the ACCC web site and non-Chinese and Chinese priests of the community have a fine command of English. Allowing for the use of English in the Chinese community in such a way helps the community gain a profile in mainstream society, because visitors are able to relate to what is happening through language and see how the community is different, but also the same in many respects. Allowing English to be used in the community serves the double purpose of accommodating Catholic Chinese children and their descendants, who are finding the community less and less appealing.

From the case studies and community history presented above, we can see an interesting dialogue of inter-culturalism created out of the examination of what makes up the 'other' from the viewpoint of Catholic Chinese. This community seems to have borrowed and appropriated into their own culture certain elements of a religion and culture that they consider foreign but important enough for there to be a need for change by inculturation or adaptation—before and after immigration to Sydney. For the purposes of this chapter, these elements are: Western chords and harmonies and the English language. I propose that, as the first inter-cultural step in music, appropriation in the community began with the introduction of the musical genre of the Western hymn.

Today, we can see all influential elements of the past and present woven into a new convention of music making, which has been achieved by Chinese Catholic community members observing the performance practice of mainstream (English-speaking) Catholicism and comparing it with what they have brought with them from abroad. Overall, there seems to be a conscious retention of older traditions in the growth of a new, modern sound—a new 'modern tradition' perhaps, in which, within the one community, we can observe many layers of ecclesiastical song that have evolved out of certain significant undercurrents that have shaped the community since its establishment in Sydney. These layers of music have the dual purpose—depending on genre or language—of asserting the community's difference from the mainstream or its ability to assimilate.

Conclusion: singing salvation

Living as Chinese, Catholic *and* Australian in suburban Sydney, members of the ACCC perform music that fulfils essential social, emotional and spiritual functions. From scholarly studies conducted on collective singing in various contexts, there is little doubt of the power that exists in singing in a group, especially in a religious context. The study presented above confirms that this is most certainly the case in the communal migrant situation of Catholic Chinese congregations in Sydney, where ecclesiastical singing is a by-product and a useful tool for religious and cultural expression through which we can observe and understand what these people as a community have accomplished since immigration. In this particular musical culture, we see reinvention, revitalisation *and* preservation in an eclectic whole towards the sustenance of a faith with the constant negotiation of a new identity through language use in ritual and song as Catholics who are Chinese, and as Chinese who are Australian.

The preservation of old forms of singing and worship with the addition of new hymns and compositions in Mandarin, Cantonese and English with an inclination towards popular styles of music show that Catholic Chinese in Sydney have arrived at a functional, hybridised compromise in keeping as much of what they can of their homeland culture while reworking and even domesticating it to the new local landscape. Perhaps the community's current protean state is best summarised in the succinct but precise observation made by Edith Lo about the time of the CCPC choir's anniversary. In her own words, it is all about 'putting culture together with faith in order to bring the Chinese people together, and the young people too' (Interview with Edith Lo, 2000). Members are able to realise in music their culture and their religion. Clearly, the future of the community is very much in the hands of the younger generation, who can choose to sustain and nurture the socio-religious faith of their forefathers or find less and less significance in associating with members of the community. By making adaptive changes in ecclesiastical song for worship, Catholic Chinese in Sydney are with each mass and service singing for their salvation in the eternal kingdom and, more pragmatically or perhaps even metaphorically, for the survival of their community on the physical plane in the Chinese diaspora.

Acknowledgments

I am indebted to the following: Dr Stephen Wild (thesis supervisor), Dr Mandy Scott (advisor), Hong Shin Chan, Paul Maclay and, most importantly, Dr Amy Chan and Alistair Noble, to whom this chapter in its current form owes its existence.

My deepest thanks extend to members of the Australian Catholic Chinese Community: Father Bonaventure Tung (OFM), Father Pascal Chang (OFM), Father Paul McGee (SSC), Joannes Chan (WSCCC President), Joseph Chow (CCPC President), Ken Lay (former WSCCC President), Ivan Ho (CCPC Choir President), Eddie and Annie Ho, Zita Ho, Adelino and Iana Chungyue, Joseph Lee, Edith and Ken Lo, Catholic Chinese Pastoral Centre Choir and Western Sydney Catholic Chinese Community Choir.

Bibliography

Anderson, Benedict 1991, *Imagined communities: Reflections on the origin and spread of nationalism*, Verso, London and New York.

Ang, Ien 2001, *On Not Speaking Chinese: Living between Asia and the West*, Routledge, London and New York.

Averill, Gage 1994, '"Mezanmi, Kouman nou ye? My friends, how are you?": musical construction of the Haitian transnation', *Diaspora*, vol. 3, no. 3, pp. 253–71.

Beall, Stephen M. 1996, 'Translation and inculturation in the Catholic Church', *Adoremus Bulletin: Society for the Renewal of Sacred Liturgy*, vol. 2, no. 6, viewed 8 September 2005, <http://www.adoremus.org/1096-Beall.html>

Catholic College Students Association 1983, *Q ng g zhàn zh róng* [*Light Songs in Worship of God's Glory*], Second edition, Catholic College Students Association, Taiwan.

Chun, Allen 1996, 'Fuck Chineseness' In *Boundary* 2 23(2): 111-138.

Cowles, Roy T. 1965, *The Cantonese Speaker's Dictionary*, University Press, Hong Kong.

Dewey, John 1916, *Democracy and Education*, Macmillan, New York.

Frith, Simon 1996, 'Music and identity', in S. Hall and P. D Gay (eds), *Questions of Cultural Identity*, Sage, London, pp. 108–27.

Hall, Stuart 1996, 'Introduction: who needs "identity"?', in S. Hall and P. D. Gay (eds), *Questions of Cultural Identity*, Sage, London, pp. 1–17.

Heilbron, J. L. 1998, 'In diaspora', in Wang Ling-chi and Wang Gungwu (eds), *The Chinese Diaspora: Selected essays*, Times Academic Press, Singapore.

Hong Kong Catholic Truth Society 1983, *X n sòng: xìn y u g ji* [*Songs of Praise from the Heart: Congregational song collection*], Second edition, Hong Kong Catholic Truth Society, Hong Kong.

Hong Kong Catholic Truth Society 1985, *Sòng n: sìn y u g ji* [*Songs of Praise: Congregational song collection*], Hong Kong Catholic Truth Society, Hong Kong.

Hosokawa, Shuhei 1998, 'Singing in a cultural enclave: karaoke in a Japanese Brazilian community', in Toru Mitsui and Shuhei Hosokawa (eds), *Karaoke Around the World: Global technology, local singing*, Routledge, London and New York, pp. 139–65.

Kelen, Christopher 2003, 'Anthems of Australia: singing complicity', *National Identities*, vol. 5, no. 2, pp. 161–77.

Kwan, Choi Wah et al. 2000, *English-Cantonese Dictionary: Cantonese in Yale romanization*, The Chinese University Press, Hong Kong.

Lau, Frederick 2005, 'Entertaining "Chineseness": Chinese singing clubs in contemporary Bangkok', *Visual Anthropology*, vol. 18, pp. 143–66.

Lum, Casey Man Kong 1996, *In Search of a Voice: Karaoke and the construction of identity in Chinese America*, Lawrence Erlbaum Associates, New Jersey.

Martin, Michael 1998, 'Pange lingua', *Theasaurus Precum Latinarum*, viewed 19 May 2006, <http://www.preces-latinae.org/thesaurus/Hymni/Pange.html>

Michael Forrester, F. 2004, 'Lions, a cardinal and two dancing dragons: valid inculturation or monkey business?', *Los Angeles Lay Catholic Mission*, March, viewed14 November 2005, <http://www.losangelesmission.com/ed/articles/2004/0403ff.htm>

Minamiki, G. 1985, *The Chinese Rites Controversy: From its beginning to modern times*, Loyola University Press, Chicago.

Mungello, D. E. et al. 1994, *The Chinese Rites Controversy: Its history and meaning*, Steyler Verlag Nettetal, Germany.

Nettl, Bruno 1983, *The Study of Ethnomusicology: Twenty-nine issues and concepts*, University of Illinois Press, Urbana-Champagne, Ill.

O'Brien, Thomas (ed.), 1982, 'Comme le prévoit', *Documents on the Liturgy, 1963–1979*, The Liturgical Press, Collegeville, pp. 284–91.

Ogawa, Hiroshi 1993, Karaoke in Japan: a sociological overview, Paper presented at the eighth International Conference on Popular Music Studies, Stockton, California.

Radano, Ronald M. and Bohlman, Philip V. (eds) 2000, *Music and the Racial Imagination*, University of Chicago Press, Chicago.

Shorter, Aylward 1988, *Toward a Theology of Inculturation*, Orbis Books, Maryknoll, NY.

Siu, Helen 1993, 'Cultural identity and politics of difference in South China', *Daedalus*, vol. 122, no. 2, pp. 19–43.

Slobin, Mark 1993, *Subcultural Sounds: Micromusics of the West*, University Press of New England, Hanover, New Hampshire.

Slobin, Mark 1994, 'Music in diaspora: the view from Euro-America', *Diaspora*, vol. 3, no. 3, pp. 243–51.

Small, Christopher 1998, *Musicking: The meanings of performing and listening*, University Press of New England, Hanover, NH.

Stokes, Martin 1994, 'Introduction: ethnicity, identity, and music', in Martin Stokes (ed.), *Ethnicity, Identities, and Music*, Berg Publishers, Oxford, pp. 1–28.

Tan Chee-beng 1988, 'Structure and change: cultural identity of the Baba of Melaka', *Bijdragen tot de Taal-, Land-en Volkenkunde*, vol. 144, nos 2–3, pp. 297–314.

The Catholic Weekly 2004, 'Our Lady: a tribute in art', *The Catholic Weekly. Special Souvenir*, vol. 2 [brochure].

Tung, Bonaventure 2006, Letter addressed to Nicholas Ng, May.

Discography

Aì G Rén Sh ng Lù [*Music Lovers' Life Journey*], 2000, Asiana Centre Association, Ashfield, Sydney.

Endnotes

[1] A Roman Catholic feast that falls traditionally on 2 November in the Gregorian calendar, after All Saints' on 1 November.

[2] An indigenous Chinese festival also known as the Ching Ming Festival (in Hong Kong) during which the graves of the departed are swept; traditionally held about 5 April (near Easter). Prayers are made with libations and food is offered.

[3] The Veneration of Ancestors (a Confucian rite) is a civil ceremony in remembrance of Confucius and family ancestors. Jesuit missionaries condoned this ritual as they wove indigenous Chinese concepts into the Roman Catholic faith in China through the methodology of inculturation. This resulted in the Edict of Toleration in 1693, which declared Christianity licit. Pope Clement XI, however, condemned the 'Chinese rites' in 1715 and ordered them to be suppressed. The ban was not lifted until the decree of 1939 published during the pontificate of Pope Pius XII, which authorised the Chinese to take part in certain Confucian rites including the 'Chinese rites' of ancestral veneration. Later, the reforms of Vatican II further encouraged the admission of native ceremonies (such as the Australian Aboriginal smoking ceremony) into church liturgy. For more information, see Minamiki (1985); Mungello et al. (1994); and Michael Forrester (2004).

[4] Notes taken during a special celebratory mass to mark the fiftieth anniversary of the ACCC and the one hundred and fiftieth anniversary of the Promulgation of the Dogma of the Immaculate Conception. The occasion saw the blessing of the statue of Our Lady of the Immaculate Conception and the launch of the Catholic China Mission Service Project by Australian Franciscan Provincial Father Stephen Bliss.

[5] A movement by overseas Catholic communities, in particular communities of Chinese Catholics and sympathisers, to raise awareness of and support the Roman Catholic Church in China, which since the Cultural Revolution (1966–76) has been oppressed by the Chinese Government.

[6] Here, it should be noted that the ACCC is one of two main organisations of Chinese Catholics within the Archdiocese of Sydney. The other is the Chinese Catholic Community Incorporated (CCC). Due to limited time and resources, the scope of my study includes only the first organisation.

[7] The topic of Latin hymnody is worthy of a discussion on its own.

[8] The edict of Vatican II inspired the methodology of 'inculturation' in twentieth-century church reform and allowed the cultural cross-fades evident in many 'ethnic' communities worldwide. By definition, inculturation is 'the creative and dynamic relationship between the Christian message and a culture or cultures' (Shorter 1988:11) and it was implemented as early as the 1600s by Jesuit missionaries in China. Inculturation encourages local forms of Christianity to emerge out of the belief that the elements of the gospel are hidden in every culture and that it is the duty of theologians, pastors and liturgists to bring this hidden light into full view (Beall 1996).

[9] Strong examples of inculturation can be found in the Chinese New Year mass of 2004 at the Cathedral of Our Lady of Angels, Los Angeles. During the procession of clergy, there were two sets of lion dancers and two dancing dragons. There was also a Chinese lantern dance and ribbon dance performed by women during the offertory. Although said mostly in the Chinese vernacular, mass was conducted in the Roman Rite with some variations allowed by the presiding Cardinal Roger Mahony with principal celebrant Bishop Ignatius Wang (Michael Forrester 2004).

[10] Mass was first celebrated in the crypt of St Mary's Cathedral but was later moved to the Asiana Centre by Father Chang (<http://www.chinese.sydney.catholic.org.au/E_Aca.htm>, viewed 7 May 2005).

[11] According to the ACCC web site, the community caters for the 'religious, social, educational and welfare activities for all people of Catholic Chinese background' (<http://www.chinese.sydney.catholic.org.au/E_Aca.htm>, viewed 7 May 2005).

[12] The numerical representation of notes used as an alternative to Western staff notation.

[13] These hymnals are used outside Sydney. On a recent fieldtrip to visit the Chinese Catholic Association of London (August 2005), I noticed the same Pew Edition hymnals as those used in the ACA/CCPC.

[14] Depending on the authoritative source consulted, there are six, seven or nine Cantonese tones. This study is based on the viewpoint of my informants as well as *The Cantonese Speaker's Dictionary* (Cowles 1965) and the *English-Cantonese Dictionary* (Kwan et al. 2000).

[15] A closer study of Roman Catholic liturgical text in Chinese can be found in my doctoral thesis, Celestial roots: the music of Sydney's Chinese, 1954–2004 (pending).

[16] At the CCPC choir's twentieth anniversary dinner in July 2005, I heard a Cantonese pop song performed with its lyrics edited to take on a slightly more liturgical meaning. This is rich in intercultural meaning and speaks of the dynamics between the expatriate Hong Kong community and their country of origin. This is, however, a case study for another paper.

[17] This is a subject with further avenues of research, as, to date, no comprehensive study on this topic exists, particularly in Australia.

8. Singing the syllables: translating spelling into music in Tibetan spelling chant

Phil Rose

Audio samples relating to this chapter are available online at:
http://epress.anu.edu.au/titles/sounds_translation_citation.html

> Ich kann das W o r t so hoch unmöglich schätzen,
> Ich muβ es anders übersetzen

(Goethe *Faust*: erster Teil)

Introduction

Every literate culture has ways of saying how the written words of its main language are made up of units of writing. In English, this is called spelling. This chapter looks at spelling in Tibetan, where it is called སྦྱོར་ཀློག་ *'sbyor klog'* (pronounced a little like 'jaw lock'). It adopts the special perspective, however, in keeping with the theme for this volume of 'sound in translation', of how the spelling is really chanted by a lama. This is called Tibetan spelling chant (henceforth TSC). Although 'sound' appears self-explanatory—the lama is, after all, creating sound when he chants—the reader will find that in TSC, sound is not a straightforward notion. First, there are two different kinds of sound involved—namely, speech sounds and musical sounds (and it is sometimes not easy to tell them apart). Second, there are many different ways at which these sounds can be looked: acoustically, for example, or historically. These differences are important for understanding TSC.

In translating from one language into another (the first sense of 'translation' that springs to mind), for example, one moves *meaning* from one place (the source of the message, be it a book, a speaker at the United Nations, or whatever) into another (the head of the listener or reader) by preserving as much of the meaning as possible while changing the linguistic form that carries it. In Euclidean geometry, translation is to move a figure or body while preserving some of its features (it must not be rotated or dilated, for example). In TSC, unlike language, it is not meaning, but *structure* that is being moved, from one domain—language—into another, music. The structure involved is that of the spelling. Just like other examples of translation, not all aspects of this structure are preserved in TSC (since languages differ not in what they *can* say but in what they *must* say; in translating from one language into another, you cannot preserve all the original meaning).

In TSC, then, spelling structure is translated into music. To what purpose? In many cultures, considerable significance attaches to being able to spell: in Anglo culture, it is still considered a mark of education and is one of the myriad methods we, as social beings, use to discriminate between ourselves. TSC is culturally significant for several reasons. First, spelling is considered a necessary preliminary to fluent reading of the Buddhist scriptures and novice monks are taught to spell by chanting the spelling of individual syllables; it is felt that chanting helps to make a difficult task—learning to spell—less onerous. It is also likely that chanting aids in remembering important aspects of the spelling. The primary significance of TSC, however, is that creating verbal energy in this way is considered a virtuous act, thus benefiting chanter and listener(s).

This chapter will thus illustrate some of the many aspects of sound involved in translating the spelling structure of classical Tibetan into music. At the same time, it will demonstrate the necessity of combining classical techniques of linguistic analysis with musical and acoustic analysis to yield a proper description of the genre. Some examples will be given of the complex mapping between the linguistic structure, which encodes the orthographic structure, and the musical structure of the chant. The description is based on Rose (2001), a detailed and fairly exhaustive analysis of a recording of a corpus of 142 orthographic syllables chanted by an acknowledged practitioner, Lama Choedak Yutok ཆོས་གྲགས་གཡུ་ཐོག, a Tibetan Buddhist lama of the *Sa skya* ས་སྐྱ tradition.

Spelling presupposes some kind of written representation of language. The top part of Figure 8.1 shows the first few words in a passage of classical Tibetan written in Tibetan script. Like most South and South-East Asian scripts, the Tibetan script developed from an Indic precursor and was adopted to write Tibetan as spoken in the sixth to ninth centuries (Beyer 1992:41; van der Kuijp 1996:372). It is probably best known in the West as the script in which the Tibetan versions of the Buddhist canon have been transmitted (Miller 1956:1). The script runs from left to right and the real text starts after the two vertical lines. Syllables are separated by dots, so this excerpt contains 18 syllables.

The fragment of script in Figure 8.1[1] is really the start of the passage, the spelling of which is chanted by Lama Choedak. I have cut out the chanted spelling of some of the syllables (those shown in grey) to keep the recording to a manageable length and to include the spelling of several different types of syllables. So that the reader can follow the chant, the bottom part of Figure 8.1 contains a representation of the speech sounds of the chanted passage. The boundaries between the spelling sentences that spell the individual syllables are marked with vertical lines; the reader should be able to hear from the lama's pausing and breathing where these boundaries occur: this is where spelling of a particular syllable ends. The boundaries internal to the spelling sentences that mark the phrases that spell different components of the syllable are also shown.

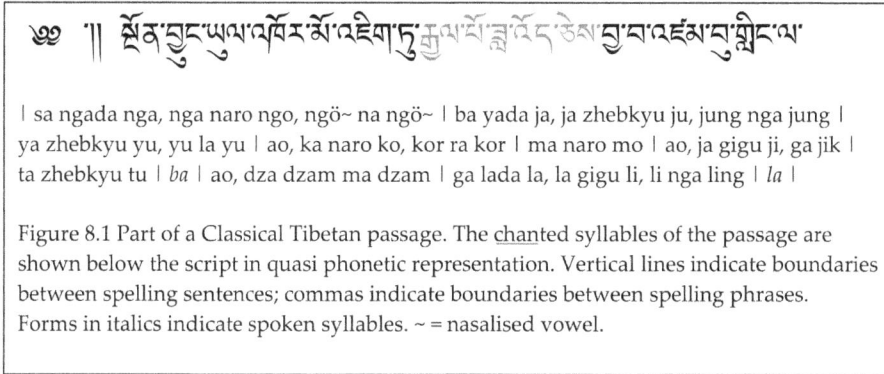

| sa ngada nga, nga naro ngo, ngö~ na ngö~ | ba yada ja, ja zhebkyu ju, jung nga jung |
ya zhebkyu yu, yu la yu | ao, ka naro ko, kor ra kor | ma naro mo | ao, ja gigu ji, ga jik |
ta zhebkyu tu | *ba* | ao, dza dzam ma dzam | ga lada la, la gigu li, li nga ling | *la* |

Figure 8.1 Part of a Classical Tibetan passage. The chanted syllables of the passage are shown below the script in quasi phonetic representation. Vertical lines indicate boundaries between spelling sentences; commas indicate boundaries between spelling phrases. Forms in italics indicate spoken syllables. ~ = nasalised vowel.

TSC is thus an expression of orthographic form in linguistic and musical structure. The remainder of this chapter will describe and exemplify aspects of the linguistic and musical structure, and show how they combine in the chant. There are two main parts to the chapter. In the first part, which deals with linguistic structure, I explain how a written Tibetan syllable such as b*Uv is spelled. In the second part, I describe aspects of the melodic and rhythmic structure of TSC and show with the help of descriptive acoustics how the linguistic structure is expressed musically, including some of the embellishments used by the chanter. Before this, however, it is necessary to discuss some interesting problems associated with the representation of speech sounds in TSC.

Representation

In Figure 8.1, the speech sounds in chant have been represented with quasi-phonetic symbols, since the use of the proper symbols, such as [ŋ], [ɔ] or [ʒ], would only introduce unnecessary complication. The reader can, after all, really hear the speech sounds involved. It should be realised, however, that the complexities in representing the speech sounds in TSC are associated not just with the use of unfamiliar phonetic symbols, but in fact go much deeper.

One complexity arises from the important distinction drawn in linguistics between two levels of speech sound structure called *phonetic* and *phonological*. A phonetic representation shows the details of the real sounds; a phonological representation shows only those aspects of the sounds that are relevant for their organisation in a particular language. For reasons that are too complex to address here—a proper account can be found in Rose (2001:166)—neither a phonetic nor a phonological representation on its own is particularly useful for the sounds of modern Tibetan in TSC.

Another complexity in representation comes from the fact that in chant the speech sounds are sung and that, in being sung, they can lose some of their linguistic features. For example, in modern Tibetan, pitch is just as much a part of a word as its vowels and consonants: it is a tone language. Pitch is, however, also manipulated in chant and the linguistic tone of a word can be modified or lost thereby. It is then a problem whether the linguistic pitch of the chanted word should still be represented. Because of these complexities, I have indicated chanted sounds in a compromise representation. Where it is necessary, I have used oblique slashes to indicate sounds in general, thus: /li/.

Linguistic structure of TSC

In much the same way as a written English word is made up of different letters, the shape of a written Tibetan syllable is made up of different components. English letters and Tibetan components have conventional names. Spelling involves saying these names using speech sounds. For example, to spell the English word *thought* we would say:

Example 8.1

/ti eɪtʃ oʊ ju dʒi eɪtʃ ti spɛlz θɔt/   "t" "h" "o" "u" "g" "h" "t" spells "thought". (1)

Here I have used the proper symbols to emphasise that the name of each of the components is itself an English word, made up of English speech sounds— for example, /eɪtʃ/ is an English word meaning 'the name of the letter "h"'. The word 'thought' has been used to demonstrate that there can often be a big difference between the spelling and the real sound. There are only three sounds in *thought*—/θ ɔ t/—but seven letters. As will be seen, this is the case also with Tibetan, and for the same reason. How words are said in a particular language—what the speech sounds are and how they go together—is referred to as phonological structure. Phonological structure is therefore a necessary component in the description of spelling.

There is, however, more to spelling than just phonological structure, since the way the names of the letters, or components, are combined—their syntax— is also part of the convention of spelling. The syntax of English spelling is relatively straightforward. In an English spelling sentence—for example, 'c' 'a' 't' spells 'cat'—the letters are first simply named in sequence and this sequence then usually constitutes the subject noun phrase: [si ɛɪ ti]NPsubj. This is followed by a verb 'spells' and then the name of the word being spelt follows in its entirety as the object noun phrase [kæt]NPobj. *Spells* and *cat* constitute a syntactic constituent called a verb phrase. This syntactic structure, which is hierarchical, is represented conventionally in linguistics with labelled brackets, thus:

Example 8.2

{[si ɛɪ ti]NPsubj. [spɛlz [kæt]NPobj.]VP }sentence   "c" "a" "t" spells "cat".   (2)

It is interesting to note from the −s on the verb *spells* that the subject noun phrase, although it consists of separate noun-like units, is considered singular, thus representing semantic rather than syntactic structure (implied is: the letters 'c' 'a' 't' *taken together as a unit* spells 'cat'. Compare this with the use of a plural verb in the sentence 'c' 'a' 't' *are* all letters of English).

Note that this is of course not the only possible spelling syntax. Different languages have different conventions. In ancient Chinese, for example, (simplifying a little) a possible spelling for a character pronounced *cat* would have been *'cat cob hat cut'*. What this stands for is: the Chinese character *cat* is spelt with the initial consonant of the character pronounced *cob* and the rhyme of the character *hat* (Malmquist 1994:10). ('Cut' stands for the Chinese name for this particular method of spelling: fǎnqiè or *reverse cutting*.) We shall see a much more complicated spelling syntax for Tibetan.

Thus to describe the linguistic structure of spelling, we must refer to phonological structure and syntactic structure. The same applies to Tibetan spelling, except that its syntax is far more complicated and interesting than English.

A simple example

Figure 8.2 The syllable *ba*.

The best way of giving an idea of how Tibetan spelling works is to start from the simplest example and then introduce more complexity. Figure 8.2 is an example of the simplest type of syllable. It is simple because it is composed of a single orthographic unit (although written with three strokes). For obvious reasons, it is convenient to represent written Tibetan in romanised form. In the Wylie romanisation used in this chapter (Wylie 1959), ༈ is romanised *ba*. It is important to realise that the romanisation does not represent the pronunciation of the component—that is, the way its name is said. The Tibetan script was adopted to represent the sounds of Tibetan in the sixth to ninth centuries. All languages change over time and the sounds of Tibetan have changed enormously since then. Since it is the modern sounds that are used to name the components of the syllable, and how the syllable itself is pronounced, it is necessary to distinguish between the romanisation, which is a representation of the orthographic components of the syllable, and their phonological representation, which represents the real sounds used in naming the components as well as the final result. This situation is similar to English spelling: because of some very complicated changes in the pronunciation of vowels, the letter 'e' is said with an /i/ vowel, as in the word 'she', not a vowel corresponding to the way it was said when the spelling was devised.

ཨ is one of a set of 30 symbols, called radicals, most of which are romanised with a sequence of a consonant plus an *a* vowel, thus: *ba*. To spell the radical *ba* on its own in TSC, it is simply spoken. (There are two instances of single, spoken, radicals in Figure 8.1, indicated by italics.) This means we have to know how the name of the component *ba* is pronounced in modern Tibetan and this means talking about linguistic (as opposed to musical) tone.

As already indicated, modern standard Tibetan is a tone language, which means that pitch is just as much a part of the words or syllables as their consonants and vowels. In the kind of Tibetan spoken by Lama Choedak, the tonology is very simple—there are just two tones: high and low (symbolised with acute and grave accents respectively). *Ba* is pronounced with a low tone, thus: /pà/ (refer to Audio 8.1a). Listen to Audio 8.1b for a sample of one with a high tone: /pá/. This high-toned example is the name for the radical *pa*, པ. It can be heard that high and low tones are realised with a falling pitch, but the low tone has in addition an initial level or slightly rising pitch component, which gives it an overall slightly convex pitch.

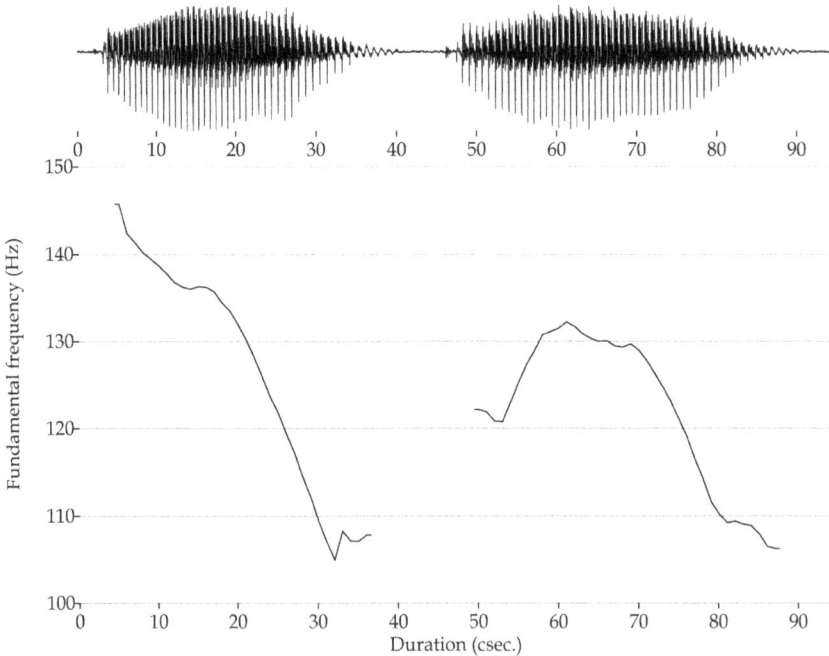

Figure 8.3 Wave-form (above) and Fundamental frequency (below) for high and low tone syllables པ /pá/ (left) and བ /pà/ (right).

Figure 8.3 shows, using the two syllables /pá/ and /pà/, the acoustics that correspond to the pitch percept of these two tones. Along the top is shown the wave form of the two syllables—the rapid fluctuations in air-pressure that is sound. It can be seen that the high tone syllable lasts for about 30-hundredths of a second and the low tone syllable is a little bit longer, at about 40 centiseconds. Below the wave form are the fundamental frequency curves for the two syllables. Fundamental frequency (F0) is the basic rate of repetition of the complex changes in air pressure. It is the primary acoustical correlation of pitch and corresponds to the rate of vibration of the speaker's vocal cords. It can be seen that the F0 shapes correspond to the pitch shapes. The F0 of the high-tone syllable falls from an onset at about 145 hertz (Hz) to about 105Hz. The F0 of the low-tone syllable rises from an onset at about 120Hz to peak about 130Hz, and then falls to the same value as the offset of the high-tone syllable.

Note that the tonal difference between the two syllables is represented in the Wylie romanisation by a difference in the initial consonant: *pa* versus *ba*. At the time the spelling was devised, it was assumed that the main difference between the two syllables was in the initial consonant. One of the major changes that Tibetan has undergone is tonogenesis, whereby contrastive linguistic tone has developed from syllable-initial consonants, and then the original difference between the consonants has been lost. This is a change found in the history of many languages. For example, the precursor of Vietnamese, a modern tone language, was not tonal. The tone of modern Tibetan is preserved only in a few well-defined cases in TSC, but, as will be demonstrated, it is sometimes reflected iconically in the musical structure, in the choice of embellishments to notes.

Vowel symbols

Figure 8.4 The syllable *bo*.

The next level of complexity is when a radical is combined with a separate symbol to indicate a vowel other than *a*. Figure 8.4 shows the syllable བོ *bo* /pò/. It can be seen that it is composed of the radical *ba* with an ⌣ symbol on top. This symbol represents the *o* vowel in *bo* and is called /nàro/ (there are three other vowel symbols, with names, representing the vowels *i, u* and *e*). The syllable *bo* is spelt /pà nàro pò/, which can be translated as 'the radical *ba* with *naro* vowel spells *bo*'. This is an example of a *vowel phrase*, a syntactic constituent that consists of three parts: the name of the radical *ba*: /pà/; the name of the *naro* vowel: /nàro/; and the result of combining them: /pò/. We represent this formally as:

Example 8.3

[pà nàro pò]vowel phrase

This example starts to give a picture of how the spelling works. An *input* is specified: /pà/; then an *operator*: /nàro/; and then the resulting *output*: /pò/—the whole being combined in one phrase. As the vowel phrase /pà nàro pò/ also constitutes a spelling sentence on its own, a more complete syntactic representation would be:

Example 8.4

{[pà nàro pò]_{vowel phrase}} spelling sentence

Postfix

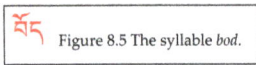

Figure 8.5 The syllable *bod*.

One of a small selection of radical symbols, called a postfix, can be added after the radical. Figure 8.5 demonstrates this. It shows the syllable *bod* (which is really the word for *Tibet*). The *ba* radical, with *naro* on top, can be easily recognised. The final symbol, the postfix, is the radical ད *da*.

Although the syllable is spelt *bod*, it is pronounced /pö\/, without a /d/ and with a vowel different from /o/: /ö/ represents a front-rounded vowel similar to the vowel in French *fleur*(flower) or German *götter* (gods) (the '\', for reasons of font incompatibility, represents the low tone). The reason for the discrepancy between the spelling and the sound is historical, like the example of tonal development mentioned above. The syllable was originally pronounced something like /bod/. The consonant /d/ is made in the front part of the mouth, just behind the teeth, by the front part of the tongue. The vowel /o/ is made with the back part of the tongue in the back part of the mouth and with the lips rounded. Historically, the originally back-rounded vowel /o/ has been pulled forwards in anticipation of the front consonant /d/, but has retained its lip rounding and thus has become a front-rounded /ö/. This is another historical change called umlaut. It is probably best known from its occurrence in the history of Germanic, where its effects can still be seen today in modern German vowel alternations in words such as *Fuß~ Füße* and their English cognates *foot ~ feet*. In the history of Tibetan, the final consonant *d* then disappeared. A low tone has also arisen from the original syllable-initial consonant *b*.

The syllable *bod* is spelt by first giving the vowel phrase to derive *bo*, then adding a postfix phrase to get *bod*, thus:

Example 8.5

{[pà nàro pò]_{vowel phrase} [pö\ tà pö\]_{postfix phrase}} spelling sentence

In Example 8.5, the vowel phrase and postfix phrase are shown combined as constituents of a spelling sentence. The vowel phrase and its structure in

Example 8.5 are clear, and it can be seen that the postfix phrase, like the vowel phrase, contains three parts. The second and third parts of the postfix phrase appear analogous to the operator and the output components of the vowel phrase, but the first part, /pö\/, needs explaining. Given the input-operator-output structure, we would expect the first term in the postfix to be the same as the result of the preceding vowel phrase—here /pò/. It is, however, one of the regularities of TSC that in the postfix phrase the input often anticipates the output. There is a possible reason for this, but it has to do with the performance of chanting itself and will be taken up again in the section below on musical structure.

Subjoined component

Figure 8.6 The syllable *byung*.

A very common component in the Tibetan syllable is found immediately under the radical. This so-called subjoined, or subfixed, component often corresponds to a *y* or *r* following the consonant spelt by the radical. In the syllables *kya* or *tra*, therefore, the *y* and *r* would be spelt by subjoined components. Figure 8.6 shows the syllable བྱུང *byung* with a subjoined *y*. The radical in *byung* is recognisable as བ *ba*, and immediately underneath it is ྱ —the subjoined *y*. Below the subjoined *y* is ུ — the symbol for the vowel *u*, called /zhèbkyu/. The postfix is ང *nga*, corresponding to the final *ng*. The subjoined portion is spelt, in a subscript phrase, with the same input-operator-output structure as the vowel and postfix phrase, thus:

Example 8.6

[pà yàda jà] *subscript phrase*

The /da/ means *joined*, so /yàda/ means '*with ya joined*', and the phrase in Example 8.6 could be translated as 'the radical *ba* with *ya* joined to it spells /jà/'. Why is the output /jà/ and not /byà/? This is because of another historical change, called labial palatalisation, whereby syllable-initial bilabial consonants such as /p/, /b/ and /m/ changed to corresponding palatal sounds such as /ch/, /j/ and /ny/ (Ohala 1978:370–3), so the syllable written *bya*, and originally said /bya/, is now said /jà/. (In some parts of Tibet, where the change did not take place, the pronunciation is still /bya/.) Labial palatalisation is also the reason why the first syllable in the name for Tibetan spelling, *sbyor klog*, is said /jo/ like 'jaw', with a palatal, but is written with a *by*. The labial palatalisation was caused by the following /y/ sound, which was also palatal, but it was an acoustically rather than an articulatorily motivated change. It happened because the acoustics of /b/ before /y/ in a syllable such as /bya/ were similar enough to the acoustics of /j/ in /ja/ for listeners to mishear /bya/ as /ja/ and then to repeat their mistake in saying what they heard—namely, /ja/.

The whole syllable *byung* is spelt with three phrases: a subscript phrase, a vowel phrase and a postfix phrase, in that order. The process is cumulative, the output of each phrase acting as the input for the next (with the exception of the postfix phrase, which, as already explained, shows phonological anticipation), thus:

Example 8.7

{[pà yàda jà] *subscript phrase* [jà zhèbkyu jù] *vowel phrase* [jùng ngà jùng] *postfix phrase*} *spelling sentence*

Further expansions

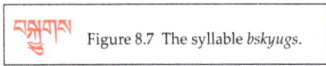

Figure 8.7 The syllable *bskyugs*.

Above I have illustrated four orthographic components in a Tibetan syllable: radical, vowel, postfix and subjoined unit. There can, however, be up to seven components. Figure 8.7 shows a Tibetan syllable with all seven: *bskyugs* (it means *vomited*). Its radical is ཀ *ka*, in addition to which a subjoined *ya* and a *zhebkyu* vowel can be recognised, making up the ཀྱུ *kyu* in bskyugs. To this a postfix ག *ga* is added, making up ཀྱུག *kyug*. The new components are: a *post-postfix* ས *sa* after the ག *ga*; a *prefix* བ, recognisable as *ba*, before the ཀྱུག *kyug*; and a *superfix* ས *sa*, on top of the ཀྱུ *kyu* complex. The syllable is pronounced /kyúʔ/ (ʔ is a glottal stop, as in cockney bu'er for *butter*, and corresponds to the postfix ག *ga*). Like silent letters in English, the three new components—the prefix *ba*, superfix *sa* and post-postfix *sa*—are not represented in the pronunciation. The syllable *bskyugs* བསྐྱུགས is spelt as follows (the /o/ is an obligatory part of the prefix phrase):

Example 8.8

{[pà o]_prefix phrase_ [sá káda ká]_superfix phrase_ [ká yàda kyá]_subjoined phrase_ [kyá zhèbkyu kyú]_vowel phrase_ [kyúʔ gà sá kyúʔ]_postfix phrase_} spelling sentence

Musical structure of TSC

One important fact about the musical structure of TSC is that much of it can be described in terms of the linguistic structure. This means that if you have to chant, say, the spelling of a vowel phrase, you know the range of possible notes on which to chant the vowel phrase components and the metrical structure in which the beats carrying them must be organised. For example, the vowel phrase of the rather complicated first syllable in the chanted passage in Figure 8.1 is /nga naro ngo/. This means adding a *naro* vowel to a *nga* radical spells /ngo/. The metrical structure assigned to such a vowel phrase in TSC would be a single measure of three crotchet-note beats (3/4), with the input /nga/, operator /naro/ and output /ngo/ components each assigned a crotchet-note value. Furthermore, the operator component /naro/ would be chanted on a double-quaver note. This is shown in the top part of Figure 8.8.

Rhythm of TSC

The acoustic basis of perceived rhythm in speech is elusive; however, the rhythm of basic beats in TSC does correlate fairly straightforwardly to acoustic duration. To demonstrate this, the bottom part of Figure 8.8 shows a so-called wide-band spectrogram of the vowel phrase /nga naro ngo/ of the first syllable. Duration, in centiseconds, is shown along the bottom and frequency, in hertz, up the right-hand side.

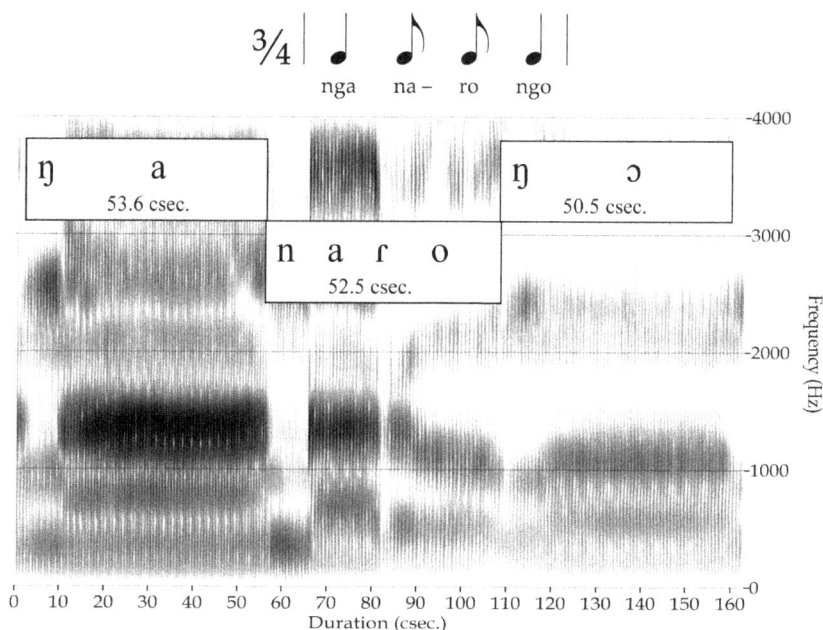

Figure 8.8 Above: Representation of rhythm of vowel phrase /nga naro ngo/of first chanted syllable. Below: Wide band spectrogram of /nga naro ngo/ showing duration of phrase constituents.

A spectrogram like this shows the distribution of acoustic energy in vocal output and how it changes over time. It shows especially the energy associated with supralaryngeal gestures such as vowels and consonants. The thick horizontal bands of energy, for example, are the supralaryngeal vocal tract resonances, called formants, and vowel quality is determined by the frequency of the first two or three of these formants (the band of energy just above 1kHz in the syllable /ngo/ is the second formant of its /o/ vowel).

A wide-band spectrogram has good enough time resolution to allow any acoustic boundaries between the units to be identified, so that the units' duration can be measured. (Quite often there are no clear-cut boundaries between speech sounds, but in this case there are.) The duration of the vowel-phrase constituents in Figure 8.8 can be measured as: 53.6 centiseconds for /nga/, 52.5 centiseconds for /naro/ and 50.5 centiseconds for /ngo/. This shows they all have very similar duration. Measurements on 39 crotchet-note beats throughout the chanted corpus (Rose 2001:178) show their mean duration to be 48.8 centiseconds (sd = 3.9 centiseconds), so the beats in Figure 8.8 are just a little longer than average. The mean beat duration of 48.8 centiseconds can be conveniently rounded off to a value of 50 centiseconds, which corresponds to a *moderato* tempo of 120 crotchet notes per minute.

Melody of TSC

In Rose (2001), five pitches were distinguished for the chant melody: B*b*, C, D, E*b* and F (B*b* is in the second octave below middle C; the others are in the next higher octave), with the lama appearing thus to be chanting in B*b*.

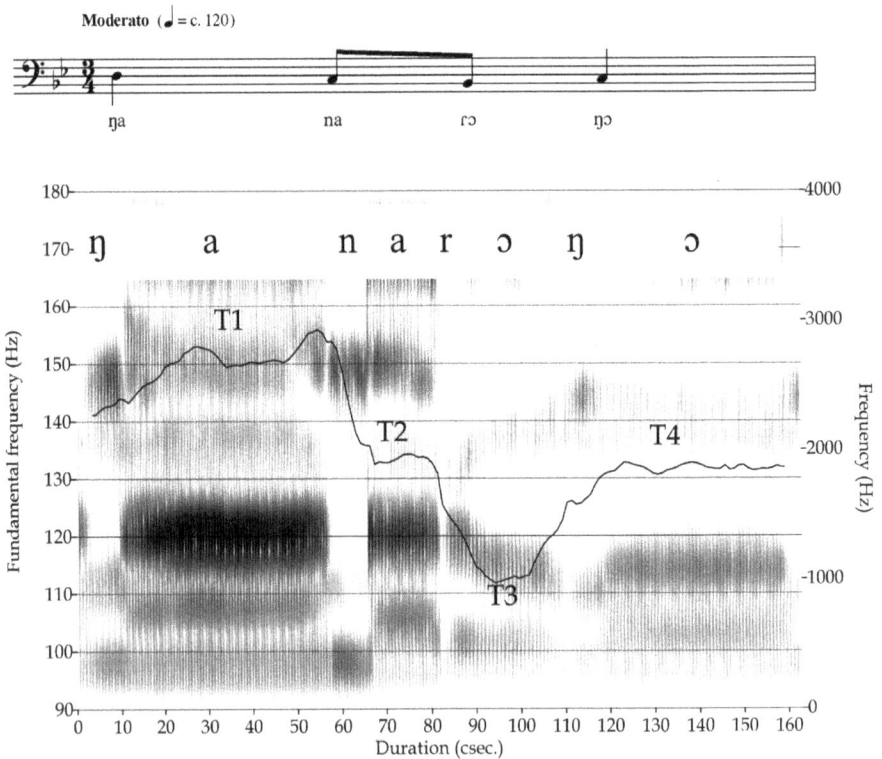

Figure 8.9 Above: musical representation of chanted vowel phrase /nga naro ngo/ D.C B*b*. C. Below: fundamental frequency of vowel phrase superimposed on its wide-band spectrogram.

Three of these pitches—D, C and B♭—can be heard on the /nga naro ngo/ vowel phrase, with its D. C B♭.C melody (in this linear representation, beats are separated by periods) (refer to Audio 8.2). Figure 8.9 shows the F0 of the vowel phrase. It has been superimposed on a wide-band spectrogram to show how the changes in F0 signalling the pitch are produced relative to the vowels and consonants of the phrase. The most important features in the F0 time course— those corresponding to my D-C-B♭-C pitch percept—are the quasi-level portions of F0 marked as T (target) 1 through 4. /nga/ D can be seen to have a quasi-level F0 target at about 150Hz. After this, the F0 falls to the second target at 133Hz on /na/ C, then falls again to the third target at about 113Hz on /ro/ B♭. Finally, the F0 rises again to the fourth target at about 132Hz on /ngo/ C, at a level, it can be seen, only slightly lower than that of the second target on /na/ C.

The F0 of speech or singling is never really static, because the vibration rate of the vocal cords producing it is affected by a myriad factors in addition to the pitch target the speaker is aiming at. The most important of these is that the vocal cords have to change their tension in moving from one target to another and their rate of vibration is also affected by the vowels and consonants they are said with.

As already mentioned, the primary acoustical correlation of pitch is fundamental frequency (F0), which is the number of times per second, or hertz (Hz), the complex sound wave repeats. In Rose (2001:183–5), I investigated to what extent the pitches of the chant melody matched their expected F0 values. In Figure 8.9, for example, it can be seen that the match is very good. The expected frequencies for B♭, C and D are 117Hz, 131Hz and 147Hz respectively (Baken 1987:487); as pointed out above, the observed values were: 113Hz ('B♭'), 133Hz and 132Hz ('C') and 150Hz ('D').

It was found in Rose (2001) that of the five melody pitches, overall, 'B♭','C' and 'F' corresponded well to the expected frequencies. There was, however, a problem with the 'D' and 'E♭' pitches, with mean F0 values of 150Hz and 152Hz (the expected values would be 147Hz and 157Hz)—although the mean F0 values corresponding to the lama's 'E♭' and 'D' pitches did not differ too much from their expected frequencies and were in fact statistically significantly different. The very small difference between them, and their bracketing by the expected F0 values, suggests that two separate pitch targets are not involved. This means, in turn, that it is not clear whether my inference above as to key ('B♭') is justified. As always, too, in cases like this where only one performance by one subject is being analysed, it is not possible to determine whether the performance is a good exemplar of the genre or whether there is a tiny lack of precision in the lama's pitch control.

In fact, the indeterminacy could be irreducible, given what is known about the relationship in professional singing between F0 and pitch targets and the way deviations from expected values are judged professionally. Sundberg et al. (1996), for example, investigated the performance of a well-known aria in commercially available recordings of 10 singers and the appraisal by professional adjudicators of the extent to which certain notes in the aria were achieved. It was found (Sundberg et al. 1996:294, 305) that the professional singers differed considerably in whether their pitch targets corresponded to expected F0 values. This means that discrepancies between pitch and F0 in this performance of chant are not necessarily indicative of inaccuracy. It was also found (Sundberg et al. 1996:295) that there was not a great deal of agreement in the professional assessors' judgments of the accuracy of the singers' notes, and further, that their judgments did not relate in a clear way to the singers' accuracy as measured by their F0. This means that it might not be possible to rely, as one must do in questions of performance correctness in phonetic fieldwork, on the judgment of other practitioners as to the quality of the chant performance.

In view of this, a reasonable interpretation of the pitch targets in TSC is that the chant melody appears to be constructed from three basic pitches—HIGH, MID and LOW—with an additional SUPERHIGH pitch of restricted occurrence. LOW corresponds to Bb, MID to C, HIGH to D or Eb in apparent free variation and SUPERHIGH to F.

Embellishments

The melody of TSC is not, of course, composed solely of plain crotchet and quaver notes. The lama makes use of several embellishments, four of which—rising and falling appoggiatura and up and down triplets—are illustrated below.

Rising and falling appoggiatura

Crotchet notes are sometimes embellished with rising or falling appoggiatura. About 10 per cent of the beats have a sharp rise in pitch onto a level target (refer to Audio 8.3). It is on the first syllable of a spelling sentence for ᷄ de, consisting of the vowel phrase /ta trengbo tè/: 'the radical da with drengbo spells /tè/' (/trèngbo/ is the name of the vowel symbol ᷄, transcribed e. For phonetic reasons, it often sounds like /ringbo/ in chant). It can be heard that the first syllable is chanted with a rising appoggiatura onto a HIGH pitch target (Eb).

Figure 8.10 shows the musical representation of the spelling sentence and the corresponding acoustics. It can be seen that the F0 onsets LOW, at about 120Hz, but rapidly rises to the HIGH target at about 152Hz by mid-syllable (compare with E♭ F0 at 156Hz). It can also be seen that the duration of the embellished beat is a little less than 40 centiseconds, showing that the embellishment is accomplished within the duration of a single beat.

Syntactically, rising appoggiaturas are associated almost exclusively with constituent-initial position, and most of these occurrences are in fact at the beginning of a spelling sentence. Phonologically, most rising appoggiaturas also occur on syllables with low tones (Figure 8.10 is an example). It is, however, also found to a limited extent on high-tone syllables with aspirated stops /ph th kh/ and voiceless fricative /s/ (Rose 2001:186). The very first syllable (*sa*) of the chant in Figure 1.2 is an example. Aspirated stops and voiceless fricatives, and Tibetan low tone, are speech sounds that are made with a relatively spread glottal aperture, and belong to Halle and Stevens' (1971) natural phonological class of so-called *spread-glottis* sounds. This suggests that rising appoggiatura has to do primarily with glottal configuration and that the otherwise nicely iconic tonal conditioning (low tones can take rising appoggiatura) is, at least to a certain extent, epiphenomenal.

The spelling sentence in Figure 8.10 also shows another typical feature of chant— namely, the spoken final syllable (this is transcribed with cross noteheads and a slur to represent the gradually falling pitch of speech; in the linear musical representation, it is indicated by italics, thus: *C Bb*). Spoken syllables have well-defined occurrence in chant (Rose 2001:179). They are obligatory in utterance-final position, in constituents such as vowel phrases, as in the example in Figure 8.10, and also when the spelling sentence consists of a single radical, of which there are two examples, indicated by italics, in Figure 1.2. In Figure 8.10, the F0 on the final syllable falls, in a way typical for speech, from 132Hz to 109Hz, and also has a duration shorter than the normal crotchet beat.

The lama also embellishes notes with falling appoggiatura. In Audio 8.4, you will hear an example in which the first syllable of the postfix phrase is /si ta si/ (*si* with postfix da spells /si/). This postfix phrase spelt the last part of the much longer syllable སྲིད *srid*. Figure 8.11 shows the musical representation and the corresponding acoustics. The first syllable is chanted with a falling appoggiatura onto a MID (C) pitch. It can be seen that the F0 starts very high, at about 160Hz, and then, over about the first half of the vowel, falls abruptly towards the MID (that is, C) target. (At least part of this initial fall is due to factors associated with the syllable-initial /s/.) The F0 overshoots the target a little, reaching about 128Hz. It then recovers to about 132Hz, almost exactly the frequency of C at 131Hz. As with rising appoggiatura, the fall is accomplished within the duration for the beat.

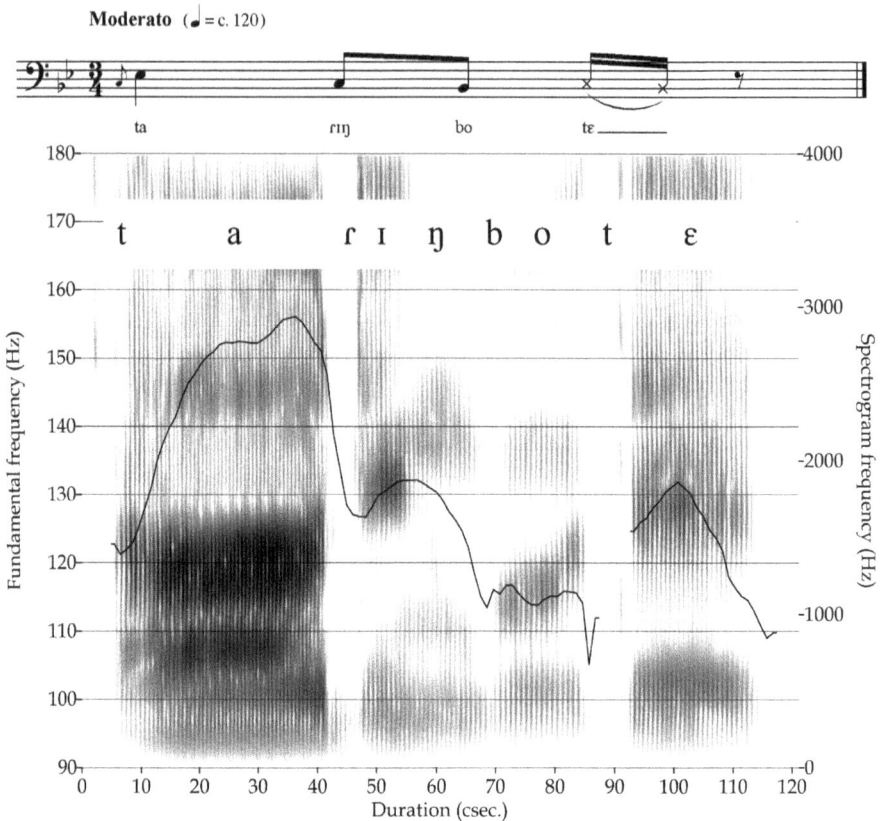

Figure 8.10 Example of rising appoggiatura in chanted spelling sentence /ta trengbo tè/ C-Eb. C Bb. C Bb. Above: Musical representation of spelling sentence. Below: fundamental frequency of /ta trengbo tè/ superimposed on its wide-band spectrogram.

The example in Figure 8.11 illustrates two more predictable features of TSC postfix phrases relating to the operator (here *da*) and the output (here the final syllable /si/). The operator in a postfix phrase is invariably given a MID (that is, C) target crotchet note. In Figure 8.11, it can seen that the F0 on /ta/ remains fairly stable throughout the vowel. Its mean value is 132Hz—again, very close to C at 131Hz. The duration of the syllable, 51.6 centiseconds, is also very close to the mean value of 50 centiseconds mentioned above for chanted crotchet notes.

Also predictable is the final syllable. It can be seen from Figure 8.11 that the final syllable /si/ is very short: its duration is 23.1 centiseconds, of which the vowel lasts 11.5 centiseconds. This is very close to the mean duration of such postfix-final syllables in the chanted corpus of 21.7 centiseconds. This duration is thus short of half the duration of the crotchet notes in the corpus (this is why it has been represented as a semiquaver note with an Abraham-Hornbostel

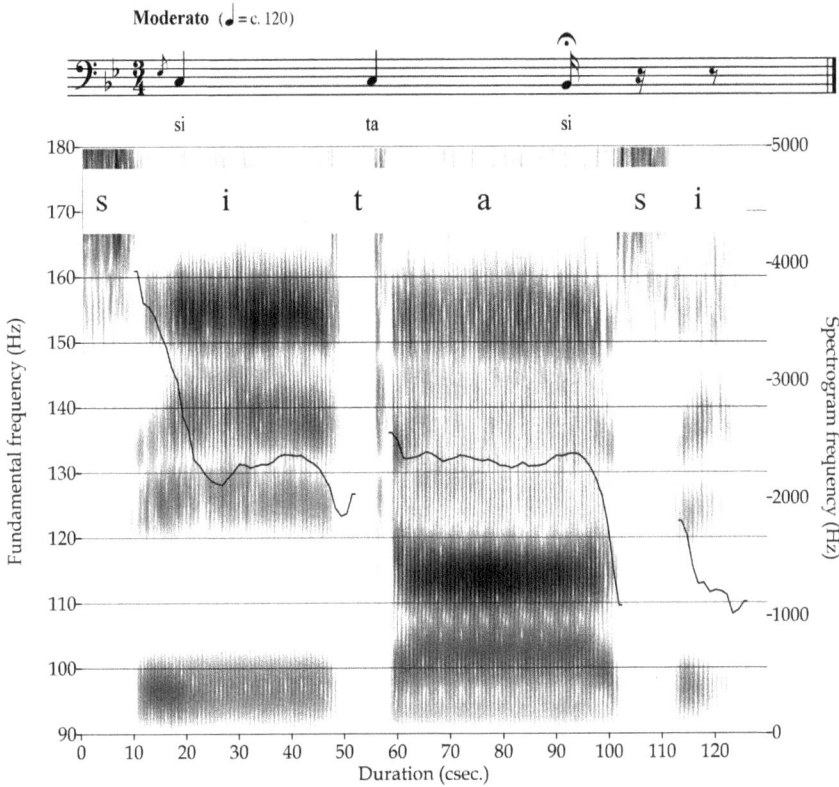

Figure 8.11 Example of falling appoggiatura. Above: musical representation of postfix phrase /si ta si/ E*b*-C. C. B*b* with falling appoggiatura on initial syllable. Below: Fundamental frequency of /si ta si/ superimposed on its wide band spectrogram.

extension symbol) (Nettl 1964:106). All postfix-final syllables have this truncated duration. The syllable also has a fully predictable LOW (B*b*)pitch. The F0 of this particular example can be seen to be at about 113Hz and is typically at a value a little lower than B*b* (116.5Hz). As with the first syllable F0, the initially falling F0 is a function of the syllable-initial /s/ and can be discounted as a pitch correlation. This extra-short B*b* note is thus fully predictable for the output of a postfix phrase.

It is possible that the very short duration of postfix-final syllables is related to the phonological feature of anticipation, mentioned above, whereby the input of a postfix phrase often anticipates its output. The anticipation is presumably because the shortness of the final syllable makes it difficult to hear what the resulting form really is, and therefore it is good to have already signalled it. Of course, this does not explain why the final syllable—the culmination of the whole spelling sentence—is so short in the first case! It could be that the shortness of the final syllable is to allow a breath to be taken without compromising the overall rhythm and delivery of the chant.

Up and down triplets

Upwards and downwards triplets constitute two more embellishments, although they are more rare than the appoggiatura. Unlike the appoggiatura, there is no phonological restriction on the type of consonant or tone that triplets occur with, but there could be some iconicity involved in their clear conditioning by syntactic position: upwards triplets are clearly associated with input position and downwards triplets with output (Rose 2001:189–90).

Refer to Audio 8.5 to hear an example of an upwards triplet taken from a spelling sentence in Figure 8.1. The triplet, from HIGH to SUPERHIGH to HIGH (E♭ to F to E♭), is on the first syllable of the vowel phrase in the spelling sentence for 𑒵 *gling*: /la kigu li/ (the vowel *gigu* with *la* spells /li/. /kigu/ is the name of the vowel symbol ꞈ, romanised *i*). The musical representation and corresponding acoustics for the phrase are shown in Figure 8.12.

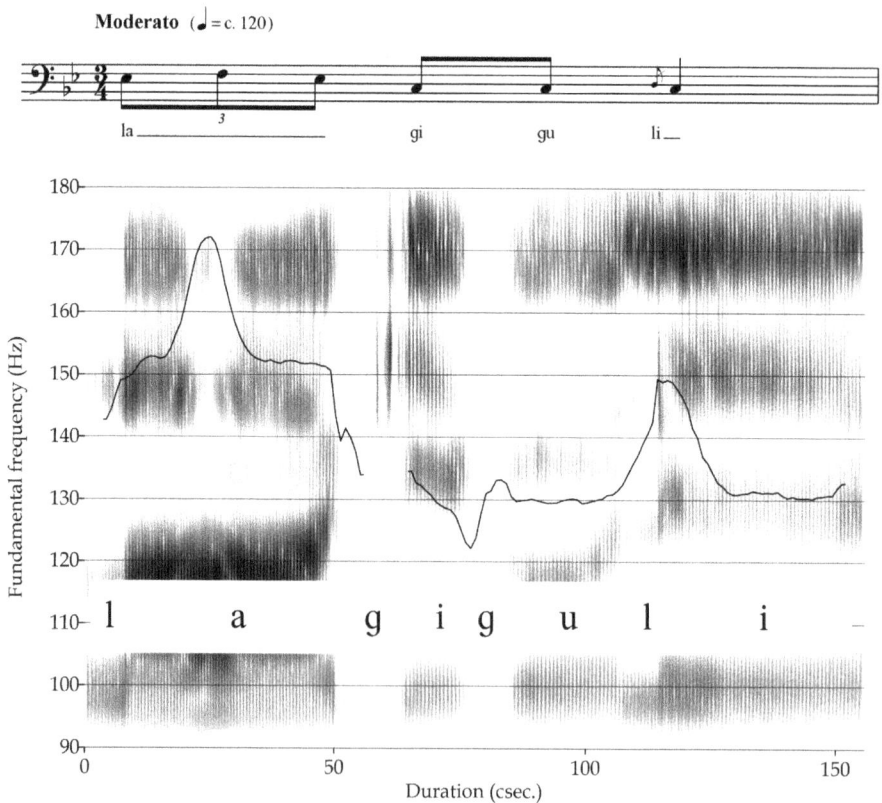

Figure 8.12 Example of upwards triplet. Fundamental frequency of the vowel phrase /la gigu li/ E♭-F-E♭.C C. D-C superimposed on its wide band spectrogram. The first syllable /la/ carries an upwards triplet.

Figure 8.12 shows that the F0 on the /la/ is at about 153Hz with a rapid mid-syllable obtrusion to about 172Hz (the rising F0 at the onset of /la/ is from a lower F0 value on the previous syllable). The expected frequencies of Eb and F are 156Hz and 175Hz respectively, so the F0 in this token is again close—within about 3Hz—to that expected from the musical pitch percept, and the centrally located obtrusion, with quasi steady-state F0 values either side, also agrees nicely with the triplet interpretation.

As far as the rest of the phrase is concerned, the remaining three beats all have a MID (C) target and their F0, at about 130Hz, can be seen to be very close to the expected 131Hz. As can be heard, the last beat has a falling appoggiatura onto C from D and the F0 corresponding to this embellishment can be easily seen. D corresponds to 147Hz, so the agreement is again very good between musical pitch and acoustics. Note that in order to produce this falling appoggiatura, the lama times its F0 fall precisely at the onset of the /i/ vowel in

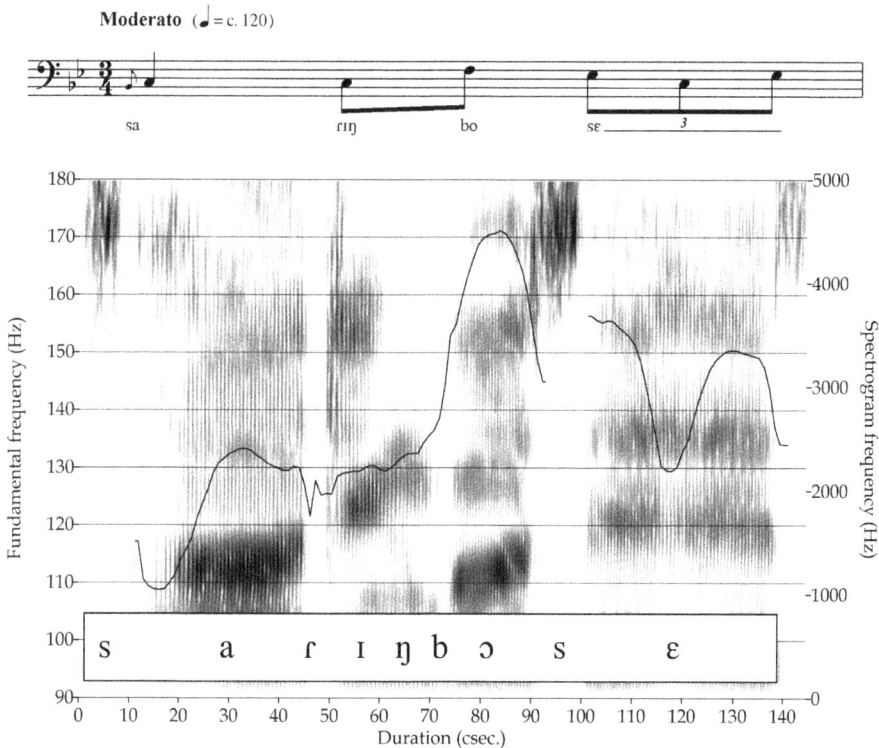

Figure 8.13 Example of downwards triplet. Above: musical representation of the vowel phrase /sa trengbo se/ Bb-C.C F. Eb-C-Eb with last syllable /se/ carrying a downwards triplet. Below: fundamental frequency of /sa trengbo se/ superimposed on its wide band spectrogram.

/li/, and times the necessary increase in F0 from the preceding C target so that it occurs during the initial consonant of the /li/. This is a typical strategy in speech whereby F0 changes that are not intended to signal pitch changes are timed to occur at the same time as syllable-initial consonants (Rose 1989:61–79).

An example of a downwards triplet, on the last syllable /se/ in the vowel phrase /sa trengbo se/, can be heard in Audio 8.6. This is from the spelling sentence for sel *sel*. The musical representation and acoustics are in Figure 8.13. It can be seen that the F0 in the last syllable vowel shows a rapid drop in its middle to about 130Hz, followed by a recovery. The F0 on either side of the drop is between about 156Hz and 150Hz, characteristic of the lama's HIGH (E*b*) pitch target.

The rest of the vowel phrase is notable for the rising appoggiatura on the initial syllable /sa/ (another case of a rising appoggiatura occurring on a syllable with a high tone, but spread-glottis initial consonant). There is also an occurrence of the SUPERHIGH (F) pitch target on the second quaver-note of the /trengbo/ vowel operator. As can be seen, the F0 reaches a peak of about 170Hz, which is near the expected frequency of 175Hz for F.

Summary

This chapter has illustrated some basic features of TSC—how the syllables of classical Tibetan are sung. In so doing, and in keeping with the theme of this volume, many aspects of sound have been invoked. Representations of sound, reflecting structure at many different levels, have been exemplified. Sound has been represented acoustically, in spectrograms and fundamental frequency curves; phonologically, in transcriptions of spelling sentences; historically, in the changes that have occurred to speech sounds still preserved in the romanisation; and perceptually, in the melody and rhythm structures of the chant. It is necessary to understand all these levels of structure to be able to describe and analyse TSC.

Another important thing demonstrated in this chapter is how TSC unites the two most important functions of sound for us—sound as speech and sound as music—so that the speech expressing the structure of the spelling is echoed and enhanced by the music. It is not surprising that speech and music relate in this way in TSC. They are both produced by the same vocal tract and are driven by the same brain (although probably not the same parts of the same brain). This is reminiscent of Pinker's (1997:535) hypothesis that 'music borrows some of its mental machinery from language'. The music of TSC, however, is not, as Pinker would have it (1997:529), the communicator of 'nothing but formless emotion'. Quite the contrary: since, as I have pointed out, most of its music is predictable from the spelling structure, the amount of variation free to convey emotion must be limited. One can, for example, infer many aspects of the linguistic structure of a spelling sentence from the rhythm and melody of its accompanying chant.

Ethnomusicologists generally agree that Western notation is theoretically descriptively inadequate for transcribing some genres of music, but nevertheless continue to use it for pragmatic reasons:

> Western notation…incorporates some of the characteristics of Western cultivated music and tends to accommodate the transcriber's subjectivity which is usually rooted in Western cultivated styles. But Western notation can be modified and, because of the facility with which it can be used, it offers the most practical method of presenting new musical data in visual form. It forms the best basis for analysis and description of music. (Nettl 1964:128)

Its use for TSC, in conjunction with the usual set of additional symbols such as grace notes (Nettl 1964:106), can, however, be justified on very much stronger grounds than just pragmatics. There is, first, the generally extremely good agreement, exemplified above, between the acoustics and my conventionally represented melodical and rhythmic percept. The syllables in a measure do indeed have durations that correspond to the notated rhythm. With the exceptions commented on in the text, the notes transcribed with a particular pitch do indeed have the expected fundamental frequency. Second, the fact that the spelling structure (as specified by the linguistic analysis) can be put into such good correspondence with the musical structure (as represented in Western notation) can also be advanced as an argument for the adequacy of the notation and the constructs it represents. Given these kinds of agreement, it would be perverse to insist that the notation is not a good model for some aspects of TSC musical structure. Just how good can be appreciated from the quality of an utterance with pitch and duration synthesised from the musical notation. Refer to Audio 8.7 to hear a version of the chant's first spelling sentence (sa ngada nga, nga naro ngo, ngö ~ na ngö ~) synthesised from the musical notation using Fujisaki's (2004) command–response model. This utterance, the middle vowel phrase of which was described above, was synthesised assuming a moderato tempo of 120 crotchet notes per minute and a pitch sequence of Bb (on sa ngada) C (nga) D (nga) C (na) Bb (ro) C (ngo) D∩C (ngö~) C (na) Bb (ngö~). Refer to Audio 8.8 to compare with the original. TSC does indeed appear to be structured largely in terms of rhythmic-melodic beats. One should not forget, also, that notation is simply a convention. One defines the convention, as I have done here, to represent as accurately as possible what are assumed to be the essential features of the performance.

There is still much to be learnt from this fascinating interaction of speech and music, not least how the results of the analysis can best be used to teach students of Tibetan Buddhism how to chant spelling authentically. The analytical approach exemplified in this chapter should, however, not blind us to the holistic significance of Tibetan spelling chant as one small part of a culture that is fighting for survival. That is, after all, where its true meaning lies.

References

Baken, R. J. 1987, *Clinical Measurement of Speech and Voice*, Collet-Hill Press, Boston.

Beyer, S. V. 1992, *The Classical Tibetan Language*, State University of New York Press, Albany.

Fujisaki, H. 2004, 'Information, prosody and modelling—with emphasis on tonal features of speech', *Procedural Speech Prosody*, vol. 1, p. 10.

Halle, M. and Stevens, K. N. 1971, 'A note on laryngeal features', *MIT Quarterly Progress Report*, vol. 101, pp. 198–213.

Malmquist, G. 1994, 'Chinese linguistics', in G. Lepschy (ed.), *History of Linguistics. Volume 1: The Eastern Traditions of Linguistics*, [English translation], Longman, London and New York, pp. 1–24.

Miller, R. 1956 [1979], 'The Tibetan system of writing', *American Council of Learned Societies Program in Oriental Languages Publications Series B— Aids—Nr. 6*, American Council of Learned Societies, Reprinted by University Microfilms International, Ann Arbor.

Nettl, Bruno 1964, *Theory and Method in Ethnomusicology*, Collier-Macmillan Ltd, London and New York.

Ohala, J. 1978, 'Southern Bantu vs. the world: the case of palatalisation of labials', *Proceedings of the Fourth Berkeley Linguistic Society Meeting*, pp. 370–86.

Pinker, S. 1997, *How the Mind Works*, Norton, New York.

Rose, P. 1989, 'On the non-equivalence of fundamental frequency and pitch in tonal description', in Bradley, Henderson and Mazaudon (eds), *Prosodic Analysis and Asian Linguistics: To honour R. K. Sprigg*, Pacific Linguistics, Canberra, pp. 55–82.

Rose, P. 2001, 'Sbyor Klog sBYor.k)og. a musical and linguistic analysis of Tibetan spelling chant', in Hannes Knifka (ed.), *Indigenous Grammar Across Cultures*, Peter Lang, Frankfurt am Main and New York, pp. 161–216.

Sundberg, J., Prame, E. and Iwarrson, J. 1996, 'Replicability and accuracy of pitch patterns in professional singers', in Davis and Fletcher (eds), *Vocal Fold Physiology–Controlling Complexity and Chaos*, Singular Publishing Group, San Diego, pp. 291–306.

van der Kuijp, L. 1996, 'The Tibetan script and derivatives', in Daniels and Bright (eds), *The World's Writing Systems*, Oxford University Press, Oxford, pp. 431–41.

Wylie, T. 1959, 'A standard system of Tibetan transcription', *Harvard Journal of Asiatic Studies*, vol. 22, pp. 261–76.

Endnotes

1. Part of a classical Tibetan passage. The chanted syllables of the passage are shown here in the script in quasi-phonetic representation. Vertical lines indicate boundaries between spelling sentences; commas indicate boundaries between spelling phrases. Forms in italics indicate spoken syllables; ~ = nasalised vowels.

9. Voice-scapes: transl(oc)ating the performed voice in ethnomusicology

Henry Johnson

At the heart of ethnomusicology is the ethnographic study of people making music. As an approach to documenting, describing, analysing and understanding the symbolic sound systems in which the world's peoples live, ethnomusicological field research includes an inherent epistemological dilemma for the researcher: the question of interpretation and how best to present and represent in scholarly discourse the music, sound and people under study. Like all studies that deal with people, sound aesthetics in all cultures will necessarily lead the researcher to encounter questions relating to interpretation and translation of data collected in the field. Whether with ethnomusicology at home or in a culture other than one's own, the intersection between field research, researcher and presentation of research findings provides a challenging space of negotiation in which processes and methods of data collection and writing up of that information are highly contested. It is from this juncture that this chapter explores the notion of voice-scapes (vocal sound-scapes performed in musical ways through speech and chant) as sites that are translated using varying modes of scholarly interpretation, and as a means of sonic communication that might be translocated into discourses that rarely study its performative and musical qualities. My aim is to problematise through ethnographic description and analysis some of the challenges that face ethnomusicologists regarding the ways sounds are perceived and how they are translated and translocated into scholarly discourses.

While it is certainly the aim of any ethnographic research to present as authentically as possible the culture, subculture or individual under study, it goes without saying that the very nature of scholarly criticism is based on the reinterpretation of data and knowledge. In the field of translation studies much has been done to make visible the processes and problems of any type of translation and, within the broad area of cultural studies, the work of Homi Bhabha, for instance, has extended the notion of translation in contemporary interdisciplinary research, especially regarding his ideas on the 'translational transnational' (Bhabha 1994:173). As a discipline now entering its sixth decade, ethnomusicology has historically faced the challenge of interpreting the music and sounds of cultures that have largely been an ethnographic 'other'. While recent trends in the broadly defined discipline have done much to introduce research at 'home', as well as being more reflexive in ontological inquiry, questions of interpretation and translation are inevitably inescapable traits of a field based on ethnographic inquiry.

The dual nature of much ethnomusicological research concerns transl(oc)ating sound, and for this chapter that sound is produced by the voice. Music of one culture is usually translated or interpreted and then relocated in another culture, or at least another cultural context, through scholarly discourse. The space that exists between musical practice and cultural translation is one that can be highly disputed and the researcher has the responsibility to present and represent as authentically as possible the culture under study. Even a 'thick description', to use Geertz's (1973) term, is one that can embody second and third-hand ethnographic information that is sometimes far removed from its original source, where the researcher negotiates an appropriate way of interpreting the symbols of another culture. It is here that the place of interpretation can be seen to occupy a liminal space, in between two clearly defined cultural realms. There is indeed a transitional state in this interpretative pathway, the outcome being defined by individual research approach and scholarly rigour.

In this chapter, I focus on the voice and the sound-scapes it produces in contexts that are usually not considered musical settings, but nevertheless maintain musical traits. The discussion includes several examples from my own ethnomusicological research as a means of challenging the ways music researchers translate concepts, whether from one culture to another or from one individual to another in the same culture. I build on the notion of 'voicescapes' (Johnson 2004; compare Smith and Dean 2003), contexts of sound production in which the chanting voice is performed as a means of human expression, and sonic spheres that challenge definitions of 'music' and 'non-music'.[1] While Smith and Dean (2003:113) use the term 'voicescapes' in a similar way to describe 'multidimensional and multidirectional projections of the voice into space, and create their own kinds of cultural geographies', in using the term, I draw more on the pioneering work of R. Murray Schafer in the field of sound-scape to describe settings in which the voice is used in musical ways yet normally considered outside the parameters of song (for example, Schafer 1977, 1992, 1993; compare Feld 1982). While Schafer does not directly make the voice a main part of his work, I extend his notion of sound-scape to one of voice-scape, in which vocal utterances create a sonic environment enmeshed in socio-cultural meaning pertaining to notions of 'sound', and subsequently challenge culturally specific definitions of the all-embracing concept 'music'. I also draw on Kuiper's (1996) work on what she terms 'smooth talkers', but extend her study to one that is related more to music research rather than linguistic/socio-linguistic analysis. Voice-scapes are part of the sound world in which we live and form meaningful sound environments through which humans communicate and express themselves through speech and chant (compare Bauman and Sherzer 1989; Ostwald 1973). They are performed sound-scapes that can have an intense effect on the listener as a result of the musical traits inherent in their articulation. They are socially positioned within a particular social context and create a sound world that is performed through a meaningful and sometimes chanting voice that commands a position that constructs musically a context that moves beyond everyday speech. Voice-scapes demand the attention of the listener, add expression to a social situation, position the chanter vis-a-vis the listener and are performed acts of social and paralinguistic communication. Such contexts include, for example, expressive speech and announcements at one end of a speech–chant continuum, and an auctioneer's chant, sports commentaries and demonstrations at the other end.

The socially performed voice challenges the boundaries of music and non-music and is an area that has not received the scholarly attention it deserves within music/sound/performance research. This chapter is an attempt to contribute to a discourse on the voice as a way of emphasising the musical qualities in its performed voice-scapes. I elaborate on this area of sound aesthetics in several cultural contexts, drawing on my own field research in different cultures. It is here that ethnographic study can raise questions regarding the translation or translocation of sound in the context of ethnomusicological research. By using case studies, I show some of the issues relating to the translation of ethnographic data.

Speech–song continuum

A premise underlying this discussion is that there is a continuum between the notions of speech and song. According to the *Oxford English Dictionary*, speech and song are defined as:

> *Speech*: The act of speaking; the natural exercise of the vocal organs; the utterance of words or sentences; oral expression of thought or feeling.

> *Song*: The act or art of singing; the result or effect of this, vocal music; that which is sung.

> (Oxford English Dictionary 2005)

The definitions of speech and song used in this discussion are based on the use of the voice in terms of its vocal production, rather than on the way it might be perceived in any one culture. That is, one type of voice production might be considered part of music making in one culture but not in another. The idea of a speech–song continuum provides a way of locating a particular type of voice production among many, where specific techniques are used towards each end. For example, in connection with the acoustics of the voice, Sundberg (2001:123–5) notes that 'the voice organ seems to be used in the same way in singing as in speech', although 'in singing…the possibilities inherent in the normal voice organ are used in quite special ways'. In singing, the voice uses such techniques as breathing, vibrato, register and pitch, whereas in speech the same techniques are used but they are not as pronounced.

In this context, the chanting voice might be discussed through an ethnomusicological approach (that is, the anthropology of sound). In doing this, one soon realises that 'certain cultures make a distinction between what is referred to as speech or talking and what is referred to as song or singing. Other cultures do not necessarily make this distinction. Other cultures distinguish forms other than speech or song which to us may seem to be intermediate forms' (List 1963:3). Rather than distinguishing between speech and song as non-music and music respectively in the way that List does (see also Merriam 1964), I show that speech and song are better thought of on a continuum that does not foreground such binary opposites, and that there are musical qualities in speech and other modes of voice production along this continuum that should be explored as part of a broader anthropology of music or sound. As Seeger (1986:59) has stressed, the separation of disciplines that study different aspects of 'vocal and verbal art has had a disastrous effect on the development of our thinking about them'.

Voice-scapes are found in many cultures. While it is sometimes problematic to call something music when its performers do not conceptualise it as such, an anthropology of music must necessarily apply outsider analysis of concepts as a way of understanding or translating some of the parameters of music-making behaviour. That is, when using the paradigm for ethnomusicology suggested by Merriam (1964), for example, which takes music sound, behaviour and conceptualisation as a basic three-part model for the study of the anthropology of music, it is the conceptualisation level that is challenged when looking at sound objects and behaviour that have musical traits.

One ethnographic example that challenges the relationship between music and speech is found in Kuna indigenous discourse in the Republic of Panama:

> The relationship between language structure and musical structure, like the relationship between speech and song, can raise issues familiar to ethnomusicologists…In this way linguistic and musical precision combine toward a descriptive understanding of the totality of Kuna ethnography of communication, from the perspective of a discourse-centered approach to both language and music. (Sherzer and Wicks 1982:163)

It is cultural traits such as these that help challenge the boundaries between speech and song in some cultures or by some researchers (compare Darnell 1989; Graham 1986; Klein 1986; Revel and Rey-Hulman 1993). The ethnomusicologist George List (1963), for example, makes a valuable contribution to music research when he explores the boundaries of speech and song, although more than four decades after his pivotal work in this area it seems that the subject is still one that is relatively understudied. What is perhaps the most striking part of List's work, however, is that while discussing speech and song as '1) vocally produced, 2) linguistically meaningful, and 3) melodic', he neglects to stress the melodically 'meaningful' aspects of the performed voice. It is these aspects of socially performed voice-scapes that are stressed in this discussion as a way of showing how the voice is musical in many ways and how people from many cultures perform the voice musically to articulate meaning through paralinguistic communication.

Towards speech

Many forms of chanting—or vocal styles very close to this—are discussed in discourses on music, particularly religious chants. When, however, voice-scapes are oriented more towards the speech end of the speech–song continuum, they seldom receive the scholarly attention in music research they deserve as performed events that have inherent musical qualities. The sonic environment produced using this mode of performed voice might not usually be conceptualised as music per se, but necessarily includes musical parameters such as pitch (see Abe 1980), vocal style, phrasing, melody and intensity. Bormann and Bormann (1972; compare Crystal 1975) have stressed the power of the voice and its ability to express non-verbal communication through vocal emphasis. While discussing the sounds of language and its vocal melodies, Bormann and Bormann note that paralinguistic communication is essential to everyday speech. There are many contexts in which the voice is performed in this way, including everyday speech, announcements and chant-like monologues.

By noting several ethnographic examples where different cultures of the world classified the voice in a variety of ways, List (1963:9) devised a classification system on the speech–song continuum that moved through speech, recitation, monotone, chant, song, intonational chant, Sprechstimme and intonational recitation. One example he noted emphasised the fact that 'the Hopi Indians of northwestern Arizona distinguish between speech, *laváyi*, song, *táwi*, and announcing [intonational chant], *tí:ngava*' (List 1963:3). List's point in mentioning this is presumably to emphasise some of the ways that different cultures classify the voice, but his classification system is devoid of any attempt to recognise the importance of the social environment in which such utterances occur. While each style might contribute to a voice-scape, it is social interaction that produces meaningful situations for the voice to be performed. It is here that performance must be emphasised as the means of giving a voice-scape its social meaning. As Enders (1990:49–50) notes, 'given the mnemonic interplay between linguistic foundations for music and musical foundations for language, it becomes unimportant to distinguish between either performative manifestation inspired by imagistic contemplation: speaking or singing, rhetoric or music'. It is when the voice is performed socially that it embodies meaning in terms of its sound aesthetics through vocal articulation. An example of the oral performance of the Kuna, for instance, includes parameters in the poetics and rhetoric of the voice during curing *rituals* that provide dramatisation (Sherzer 1986:175). The use of the voice in this context is one that helps emphasise its performative aspects in certain social situations.

Another example of vocal classification noted by List (1963:5) is of three types of 'sound communication' of the Maori of New Zealand. He points out that Maori differentiate speech (*koorero*), ritual chant (*karakia*) and song (*waiata*). In his discussion, List (1963:11) compares the vocal style of Sprechstimme to that used in the *haka* (warrior dance with chant) or, as McLean (1996:44) describes it, 'a posture dance with shouted accompaniment'. Also in connection to Maori vocal styles, *waiata*, according to McLean (1996:110), is a sung style that contrasts with recitation. Furthermore, *karakia* (prayer/lament), of which there are many kinds, might be called spells or incantations (McLean 1996:35–7) that 'are performed usually in a rapid monotone punctuated by sustained notes and descending glides at the end of phrases'.

The use of the term 'everyday speech' is particularly difficult to define. Social contexts of voice production necessarily perform the voice, but are not usually perceived to be performance events, yet they are inherently performative acts that are played out socially in just about every encounter. A voice-scape towards the speech end of the speech–song continuum might be a conversation, a speech, a sermon, an announcement or a commentary. In each, however, there is a sonic sound-scape producing meaning through vocal articulation in a musical way, not only through linguistic meaning. This might be termed the musical voice. As noted by Gerson-Kiwi (1980:29) in connection with a study of sound alienation in Asian religious rituals, 'although not generally observed, every single one of us in the course of ordinary speech runs through a broad range of melodic curves or rhetorical intonations accompanying dry speech, and dividing, for the sake of clarity, its syntactical parts'. These musical traits, especially the melodic curves and rhetorical intonations, are evident in speech phrasing, timbre and tone (compare Crystal 1975), but it is the performed part of the context that helps give it meaning during human interaction or communication. A Japanese Buddhist ritual is a typical example. There is often a 'grey' area between speech and chant or the two are mixed. There can also be vocables or guttural sounds interpolated in the speech and/or chant. In connection with Japan, for instance, one is reminded of such problems of definition when one considers the diverse behaviour that is connected with sound in culture. The Japanese do have concepts for music: the most immediate is their adoption of the English term 'music' into the Japanese word '*myûjikku*'. Other terms, however, exist that are related to the sound, concepts and behaviour that encapsulate *myûjikku* behaviour (Johnson 1999). As Coaldrake (1989:73) has said:

An answer to the fundamental question 'What is music?' is as essential to the appreciation of the aural phenomenon that exists in Japan as it is to the understanding of patterns of organised sound in Western culture, and just as difficult to find. The difficulties are epitomized by the fact that the authoritative *New Grove Dictionary of Music and Musicians* does not even have an entry for 'music'.

During ethnographic field research in Japan, while exploring a multitude of sound-scapes that challenge the notion of 'music', I became particularly aware of the way the voice was used during everyday spoken situations. Moreover, it was the ways that the spoken voice was performed in some situations that were aurally striking, especially when music traits would be seemingly exaggerated in some social contexts so as to communicate with another person. In Japan, there are many contexts in which one can hear the chanting voice, to use a more general definition of the term. For example, chanting can be found at Buddhist temples and Shinto shrines, where there is a clear melodic line. What is interesting, however, in even these contexts is that from an insider's perspective the chanting is not always perceived as music. In contrast to these chants, in everyday speech the voice is performed in a creative way in such places as a conversation between two people. For example, a young female company employee might answer the phone to a customer by using a voice that is noticeably raised in pitch in comparison with an everyday voice. Such is the extent of this raising of pitch that the speech is almost at a falsetto level. The employee might seem almost apologetic while talking to the customer, using melody and expression as a way of socially lowering herself and raising the status of the caller (a different vocabulary would be part of such a conversation). The conversation produces an aesthetic sound-scape, especially on the side of the employee, in which the voice is performed in a way in which it is altered significantly from most other social utterances.

Announcements, too, provide examples of a near-chanting voice. 'Mind the gap' or 'The next train on platform…' are typical statements at train stations that provide a voice that is intended to be noticed, is slightly exaggerated and gives a sense of authority. Television advertisements are replete with similar announcements, ones that create a voice-scape of exaggerated speech with more expression in tempo, intonation and phrasing: 'Up to 50 per cent off!', 'You won't get a better bargain anywhere else!'. On a plane one might be asked which choice of meal one wants in an everyday voice, but when the cabin crew walks down the aisle asking who wants tea or coffee the voice is usually raised in pitch and much louder and melodic than in other contexts. Even sellers adapt their spoken voice as a tool of their trade. While discussing the differences between street language and that used on the stage, Clidière (1993:138) notes that 'all sorts of entertaining speech can be found in the town: salesmen with their smooth talk, the voices of the preacher or the propagandist, the simple drunkard insulting passerby [sic]'. Indeed, just about any announcement, anywhere, provides a sound environment in which the voice is performed through social interaction and manipulation of its musical parameters so as to capture the attention of a listener. It is primarily the musical traits of such performative acts that allow the voice to function in this way; and it is this social situation that allows the voice to be performed musically in such settings. As a means of performed expression, the spoken voice has long been used by composers, poets and speakers alike as a musical instrument with the aim of emphasising the musical qualities of speech.

The vocal style of the American civil rights activist and preacher Martin Luther King, Jr (1929–68), is one that is replete with musical qualities. The 'I have a dream' speech delivered by King on the steps of the Lincoln Memorial in Washington, DC, on 28 August 1963 is one of many of his speeches that has a vocal style that is full of such features. His style of speech was one that was performed through a theatricality of vocal presentation using a musical voice that was almost chanting to an audience. For example:

> I have a dream today.
>
> I have a dream that one day the state of Alabama, whose governor's lips are presently dripping with the words of interposition and nullification, will be transformed into a situation where little black boys and black girls will be able to join hands with little white boys and white girls and walk together as sisters and brothers.
>
> I have a dream today. (Clayton 1968)

This type of vocal performance is similar to a preaching style, especially that found in the evangelical church. Other spheres where the voice is performed in a similar way include politics (that is, speeches) and the armed forces (that is, orders).

Towards song

Moving along the speech–song continuum towards the song end, there are numerous contexts that display the voice using a chanting version of vocal production. This type of vocal technique is not speech and not song, but it is has more pronounced musical parameters.

While the area of announcements has already been mentioned, the topic can be extended here to show other voice-scapes that move more towards the sung style of the chanting voice. Walking through a market, one will often hear stallholders chanting what they have to sell. They seldom 'speak' the offer they have to sell, but usually 'chant' it. This chanted mode of vocal performance tends to exaggerate the words and include more pronounced musical parameters in the chant. Newspaper vendors too often belt their voice in a style somewhat similar to the characteristic vocal style of Martin Luther King. For example, stallholders might chant the name of the newspaper they have for sale: '*Evening Standard...Standard*', '*Star...Star*' or '*The Big Issue*, sir'. With '*Star*', for instance, the 'ar' might be stretched considerably and the final 'r' might rise in pitch and then cut off abruptly.

Another context of a similar type of chanting voice occurs in sports commentaries. A race commentator, for example, often accompanies a horserace. When the speaker begins to comment on the race, the voice is normally in a spoken style. As the race develops, however, the commentator changes the vocal style into one that is almost chant-like. The voice intensifies, it increases speed and it rises in pitch:

> Racing commentaries in play-by-play mode are droned or chanted, that is, they are basically articulated in a monotone. The intonational note usually rises in semitones to a high point at the finishing post and then gradually comes down as the commentator moves through the last cycle. On the way down the commentator also moves out of this mode back into color mode with normal speech intonation. (Kuiper 1996:19)

The commentary is a performance that not only serves a function of letting the listener know race details, it creates suspense and excitement for an audience through voice manipulation.

The context of an auction too is one in which an auctioneer uses a chanting voice to sell items at the event. As List notes:

> Distinctions made according to function rather than melodic type are also common in our own society. 'Auctioneering', the form of communication used by the auctioneer, is not usually considered 'song' or even 'chant'. Nevertheless, 'auctioneering' often takes the form of a monotonic chant in which the monotone and the few auxiliary tones used are quite stable. On occasion, types of melodic cadences are also used.
>
> When speech is heightened in a socially structured situation, such as a dramatic production or in the telling of a tale, two opposite tendencies appear. The first is the negation or the levelling out of intonation into a plateau approaching a monotone. The second is the amplification or exaggeration of intonation, especially of the downward inflection that serves in most languages as a phrase, sentence, or paragraph final. (List 1963:6)

The vocal technique used by the auctioneer is known as the 'auctioneer's voice' (see Kuiper 1996). The technique is quite striking, especially with the auctioneer's 'singsong melody' (Proffitt 2005; compare Herzog 1934), and the event might be viewed as a performance that includes the practical function of the auctioneer keeping the sale moving while displaying verbal skills and sometimes including humour. 'Some auctioneers chant very fast—relatively speaking in a monotone manner and others use varying speeds of vocal delivery and pitches in tone…When performed by an experienced auctioneer, one often hears the chant described as a melodious song or yodel' (Ector 2005). The reasons for the use of this chant-like performance are explained as follows:

> First, auctioneers have a lot to sell in a limited amount of time…Second, forcing bidders to make fast decisions is part of the strategy of auction marketing…Third, a good chant makes the auction process interesting for the bidders. The chant is the auctioneer's calling card, and the good bid callers blend art and entertainment. (Proffitt 2005)

The content of the chant contains the barest of information: 'bid prices, ask prices, and filler words' (Ector 2005). For example, an auction in the United States might go like this:

> 1 dollar bid, now 2,
> now 2, will ya give me 2?
> 2 dollar bid, now 3,
> now 3, will ya give me 3?3 dollar bid, now 4,
> now 4, will ya give me 4?

> (National Auctioneers Association 2005)

In terms of the musical content of the chant, 'many people think auctioneers sound like they're singing because the chant's rhythm has a beat much like music does. The steady rhythm allows the auctioneer's chant to move more rapidly than normal speech' (National Auctioneers Association 2005). The chanting is, however, also part of the performance, especially in the way the auctioneer carries out their job. 'Besides keeping the auction moving, the fast-paced chant creates excitement and makes the auction environment entertaining' (National Auctioneers Association 2005).

There are indeed many other contexts in which similar chanting voices are used in musical ways (for example, protests, sports supporters, and so on). Many of these contexts produce voice-scapes that are chant-like and contain musical parameters such as call and response, rhythm, meter, pitch and melody. For example, when a group of protesting workers is marching along a street, their cry might be as follows: 'What do we want? More pay! When do we want it? Now!' Such a chant would be in duple time (each phrase occupying one bar) and contain call and response as a technique of communicating an idea clearly to an audience.

The use of the (non-singing) voice in musical composition is a technique that has often been used by Western art-music composers. Whether a speaking voice, speech-song or a chanting voice, the voice is used in ways that distinguish it from the singing voice. The American composer Steve Reich (b. 1936), for example, produced the orchestral work *City Life* (1994) based on the sound-scapes of New York that he heard in the late twentieth century. In this work, Reich makes use of the voice of a street seller and builds around the chant in terms of sampling and transforming its sound. In Waffender's (1995) film of the making of this work, a street seller is clearly seen chanting 'Check it out…Every item $10, check it out…Excuse me young lady, check it out'. The last phrase is extremely exaggerated with an extended 'u' in 'excuse' and 'e' in 'check'. Reich builds on this chanting in a section of his work around the melody of the chanted phrase 'Check it out'.

Some of Reich's earlier works also use the spoken voice as a building block for his compositions. *It's Gonna Rain* (1966) uses a preacher's voice and *Come Out* (1966) uses the single phrase 'Come out to show them'. In *Come Out*, the voice of a survivor of a race riot is used and manipulated through different channels. Also, *The Cave* (1993) is one of Reich's operas that uses spoken words of interviews with Israeli, Palestinian and American people on the history of The Cave at Hebron. An interview by Andrew Ford with Reich helps show the composer's ideas while using the spoken voice as a structuring tool in his compositions:

> *Andrew Ford*: I suppose it was really 'Different Trains' was it, where you returned to the spoken voice?

> *Steve Reich*: Yes, well basically what happened was that after doing 'It's Gonna Rain' and then 'Come Out' in 1966, I felt I didn't want to sit around here making tapes, I'm a composer, I want to write live music, and I made that transference from the tape medium to the live music, and basically didn't look back. The only electronics I used between 1967 and 1988 were microphones to amplify the ensemble. In 1988 I was asked by Betty Freeman to write a piece…for the Kronos Quartet, and I became aware in pop music of the sampling keyboard. The sampling device is basically a computer that records sound as a keyboard interface, so that you can play what you record by playing keys on the keyboard. This was like an IBO kind of technology, which I had thought, was just made for me personally. It didn't pretend to imitate a live instrument the way synthesisers will imitate a trombone or a violin. Basically you bring it home from the store, you plug it into the wall, you press middle-C and nothing happens. You have to record something into it and the possibility of bringing in things from the outside world into music of this time, not with tape but something that could be played just by playing on the third beat of the fourth measure, was exceedingly attractive. I felt that I wanted to go back to these tape pieces, but now instead of as if they

were music, they could literally become part of the music. So in 'Different Trains', which was a sort of line in the same kind of piece, every time you hear a woman's voice [it's] doubled by the viola, and every time you hear a man's voice, it's doubled by the cello, and the vocal melodies, the speech melodies are the melodic cells that generate the entire piece. So it was finally a kind of fusion of bringing back the old ideas from the '60s along with tape pieces, and saying, OK, now they will serve as integral parts of a piece of live music. (Ford 2003)

In another interview, Reich further illustrates some of these points:

Weidenbaum: I was co-editing a magazine called *Classical Pulse!* at the time and K. Robert Schwarz did a story for us about that. But I have spoken with Scott Johnson about his I.F. Stone stuff, which really affected the way I heard voices. Did your work similarly affect you the way you yourself heard people speak?

Reich: Sometimes yes, but basically there are languages in the world, that we don't speak, you and I, but in Africa for instance, where—they're called tonal languages—if you don't have the melody right, then you don't have the meaning right. But even in English, American English, if someone [says] 'no', 'No!', or, 'erm, no', those are three different statements. We are used to living with what I would call 'speech melody' hovering over everything we say—it's happening right now, and it's the emotional water [*starts to laugh*] in which our words swim. (Weidenbaum 2004)

The last example illustrated here regarding the blurred boundaries between speech and song is found in Tibetan Buddhist spelling chant. Rose (2001), for example, discusses this type of chant in terms of its musical qualities, especially that relating to structure (for example, rhythm, metre and pitch):

The musical structure provides for a small choice of rhythmic-melodic units at each beat. These units make use of three basic pitch levels—high, mid, and low—with an additional super high pitch that is only found in high falls and very rarely in eighth-note pairs. The choice of rhythmic-melodic unit is restricted to a large extent by the syntactic and phonological properties of the spelling sentence, especially the tonal structure of syllables and position within a phrase. (Rose 2001:203)

In many cultures, prayer involves chanting, a behaviour that has inherent musical traits, although its practitioners do not always conceptualise their behaviour as musical.

Closing thoughts

Humans structure sounds in many ways. The study of music sound per se will not help in the understanding of the sociological processes involved in its performance, nor will it further the understanding of music as culture. While Merriam's (1964) paradigm for ethnomusicology includes the analysis on three interrelated levels of music, concepts and behaviour, there is inherent in this model the problem of establishing exactly what music is or is not. Forgetting the problems with terminology and the cultural bias that one has concerning the boundaries in one's own culture of concepts such as 'music' and 'sound', I am arguing here that there are universal forms of human behaviour where sound can be placed within an area of culture for analysis as humanly organised sound—the idea of voice-scapes helps illustrate this point.

The notion of transl(oc)ating sound is at the heart of ethnomusicological method. Reflecting on the ways music travels, whether within its own culture or interpreted and presented through scholarly discourse to other cultures, the thoughts presented in this study have been intended to problematise some of the practices that are central to ethnomusicological study: the translation and relocation of sound (in this case, sound produced by the voice). Interpretation and scholarly rigour are of course important factors in this process, ones that seek best to present and represent the music and people under study. This chapter has shown some of the grey areas of acoustic sound-scapes that are not always considered music, yet have inherent musical and performative traits that allow them to be considered in discourses on music. The discussion has shown, therefore, that there are many sound-scapes and voice-scapes alike that are still to be explored within music research. An understanding of these 'scapes' will help ultimately in the understanding of the ways people make music within and between many sound environments.

References

Abe, Isamu 1980, 'How vocal pitch works', in Linda R. Waugh and C. H. van Schooneveld (eds), *The Melody of Language*, University Park Press, Baltimore, pp. 1–24.

Bauman, Richard and Sherzer, Joel (eds) 1989, *Explorations in the Ethnography of Speaking*, Second edition, Cambridge University Press, Cambridge.

Bhabha, Homi K. 1994, *The Location of Culture*, Routledge, London.

Bormann, Ernest G. and Bormann, Nancy C. 1972, *Speech Communication: An interpersonal approach*, Harper & Row, New York.

Clayton, Ed 1968, *Martin Luther King: The peaceful warrior*, Illustrated by David Hodges, Third edition, Prentice-Hall, Englewood Cliffs, NJ.

Clidière, Sylvie 1993, 'Voix de traverse: Parole et spectacles de rue', in Nicole Revel and Diana Rey-Hulman (eds), *Pour une Anthropologie des Voix*, L'Harmattan, Paris, pp. 197–207.

Coaldrake, Kimi 1989, 'Breaking the sound barrier: the inner world of Japaense musi', *Miscellanea Musicologica*, vol. 16, pp. 71–8.

Crystal, David 1975, *The English Tone of Voice*, Edward Arnold, London.

Darnell, Regna 1989, 'Correlates of Cree narrative performance', in Richard Bauman and Joel Sherzer (eds), *Explorations in the Ethnography of Speaking*, Second edition, Cambridge University Press, Cambridge, pp. 315–36.

Ector, Cedric J. 2005, *Understanding the Auctioneer's Chant*, viewed 26 November 2005, <http://www.elac-llc.com/html/auction-chant.html>

Enders, Jody 1990, 'Visions with voices: the rhetoric of memory and music in liturgical drama', *Comparative Drama*, vol. 24, no. 1, pp. 34–54.

Feld, Steven 1982, *Sound and Sentiment: Birds, weeping, poetics, and song in Kaluli expression*, University of Pennsylvania Press, Philadelphia.

Ford, Andrew 2003, 'Interview with Steve Reich', *Radio National*, 8 February 2003, viewed 26 November 2003, <http://www.abc.net.au/rn/music/mshow/s780640.htm>

Geertz, Clifford 1973, *The Interpretation of Cultures*, Hutchinson of London, London.

Gerson-Kiwi, Edith 1980, 'Melodic patterns in Asiatic rituals: the quest for sound alienation', *Israel Studies in Musicology*, vol. 2, pp. 27–31.

Graham, Laura 1986, 'Three modes of Shavante vocal expression: wailing, collective singing, and political oratory', in Joel Sherzer and Greg Urban (eds), *Native South American Discourse*, Mouton de Gruyter, Berlin, pp. 83–118.

Herzog, G. 1934, 'Speech melody and primitive music', *Musical Quarterly*, vol. 20, pp. 452–66.

Johnson, Henry 1999, 'The sounds of *Myûjikku*: an exploration of concepts and classifications in Japanese sound aesthetics', *Journal of Musicological Research*, vol. 18, no. 4, pp. 291–306.

Johnson, Henry 2004, 'Voicescapes: the (en)chanting voice and its performance soundscapes', *Soundscape: The Journal of Acoustic Ecology*, vol. 5, no. 2, pp. 79–98.

Klein, Harriet E. Manelis 1986, 'Styles of Toba discourse', in Joel Sherzer and Greg Urban (eds), *Native South American Discourse*, Mouton de Gruyter, Berlin, pp. 213–35.

Kuiper, Koenraad 1996, *Smooth Talkers: The linguistic performance of auctioneers and sportscasters*, L. Erlbaum Associates, Mahwah, NJ.

List, George 1963, 'The boundaries of speech and song', *Ethnomusicology*, vol. 7, pp. 1–16.

McLean, Mervyn 1996, *Maori Music*, Auckland University Press, Auckland.

Merriam, Alan P. 1964, *The Anthropology of Music*, Northwestern University Press, Evanston, Ill.

National Auctioneers Association 2005, *All About the Auctioneer's Chant*, viewed 26 November 2005, <http://www.msaa.org/chant.php>

Ostwald, Peter F. 1973, *The Semiotics of Human Sound*, Mouton, The Hague.

Oxford English Dictionary 2005, *Oxford English Dictionary*, [Online edition], viewed 26 November 2005, <http://dictionary.oed.com/>

Proffitt, Steve 2005, 'The chant—real or illusion?', viewed 26 November 2005, <http://www.maineantiquedigest.com/articles/ethi0300.htm>

Revel, Nicole and Rey-Hulman, Diana (eds) 1993, *Pour une Anthropologie des Voix*, L'Harmattan, Paris.

Rose, Philip J. 2001, 'Sbyor Klog—a musical and linguistic description of Tibetan spelling chant', in Hannes Kniffka (ed.), *Indigenous Grammar Across Cultures*, Peter Lang, New York, pp. 161–216.

Schafer, R. Murray 1977, *The Tuning of the World*, McClelland and Stewart, Toronto.

Schafer, R. Murray 1992, 'Music, non-music and the soundscape', in John Paynter, Tim Howell, Richard Orton and Peter Seymour (eds), *Companion to Contemporary Musical Thought. Volume 1*, Routledge, London, pp. 34–45.

Schafer, R. Murray 1993, *Voices of Tyranny: Temples of silence*, Arcana, Ontario.

Seeger, Anthony 1986, 'Oratory is spoken, myth is told, and song is sung, but they are all music to my ears', in Joel Sherzer and Greg Urban (eds), *Native South American Discourse*, Mouton de Gruyter, Berlin, pp. 59–82.

Sherzer, Joel 1986, 'The report of a Kuna curing specialist: the poetics and rhetoric of an oral performance', in Joel Sherzer and Greg Urban (eds), *Native South American Discourse*, Mouton de Gruyter, Berlin, pp. 169–212.

Sherzer, Joel and Wicks, Sammie Ann 1982, 'The intersection of music and language in Kuna discourse', *Latin American Music Review*, vol. 3, no. 2, pp. 147–64.

Smith, Hazel and Dean, Roger T. 2003, 'Voicescapes and sonic structures in the creation of sound technodrama', *Performance Research*, vol. 8, no. 1, pp. 112–23.

Sundberg, Johan 2001, 'The voice', in Stanley Sadie (ed.) and John Tyrrell (ex. ed.), *The New Grove Dictionary of Music and Musicians. Volume 1*, Second edition, Grove, London, pp. 120–5.

Waffender, Manfred (writer and director) 1995, *City Life*, ZDF.

Weidenbaum, Marc 2004, 'Interview with Steve Reich', viewed 26 November 2005, <http://www.disquiet.com/stevereich-script.html>

Endnotes

[1] In this discussion, I use the term 'chant' according to the *Oxford English Dictionary* (Online edition): 'A singing intonation or modulation of the voice in speech; a distinctive intonation.'

www.ingramcontent.com/pod-product-compliance
Lightning Source LLC
Chambersburg PA
CBHW061245270326
41928CB00041B/3434

* 9 7 8 1 9 2 1 5 3 6 5 4 0 *